TIME
WARS

TIME WARS

The Primary Conflict in Human History

JEREMY RIFKIN

Henry Holt and Company · New York

To my parents
Vivette and Milton
and my sisters
Martyl
Dovie
Jerelyn

Copyright © 1987 by Jeremy Rifkin
All rights reserved, including the right to reproduce this
book or portions thereof in any form.
Published by Henry Holt and Company, Inc.,
521 Fifth Avenue, New York, New York 10175.
Distributed in Canada by Fitzhenry and Whiteside Limited,
195 Allstate Parkway, Markham, Ontario L3R 4T8.

Library of Congress Cataloging-in-Publication Data
Rifkin, Jeremy.
Time wars.
Bibliography: p.
Includes index.
1. Time—Philosophy. 2. Time—Social aspects.
3. Time—Political aspects. 4. Time perception.
I. Title.
QB209.R54   1986   303.4   86–25845
ISBN: 0-8050-0377-0

First Edition

Designer: Victoria Hartman
Printed in the United States of America
1 3 5 7 9 10 8 6 4 2

ISBN 0-8050-0377-0

Contents

V | The Democratization of Time   189

TIME
WARS

INTRODUCTION

Time is fundamental. It is the principle that underlies and permeates our physical and biological systems. It is the language of the mind, informing our behavior and defining our personality. Time is the instrument that makes possible group interaction and the creation of culture.

The temporal realm stretches out to the far corners of the universe and eases its way down into the smallest structures of subatomic life. Of all the symbolic forms the human family has invented, time is the most inclusive. Time is our window on the world. With time we create order and shape the kind of world we live in. Yet we take our time values for granted, never stopping to consider the critical role they play in defining the social order.

Every culture has its own unique set of temporal fingerprints. To know a people is to know the time values they live by. To know ourselves, why we act on each other and the world the way we do, we must first understand the temporal dynamics that govern the human journey in history.

Homo sapiens is, in the words of the scientist Alfred Korzybski, the only "time-binding" animal. All of our perceptions of self and world are mediated by the way we imagine, explain, use, and implement time. Time is at once both dazzling and versatile, enigmatic and vexing. We can look ahead of ourselves, we can steal our way back into the past, we can detach ourselves from the moment and look at ourselves from a distance. Our clocks and schedules, our science and technology, allow us to leap on top of the undifferentiated tempos of the biological and physical world. We ride herd on the periodicities of nature. We tame, harness, and regiment. We brand our temporal biases onto the ancient rhythms of the universe, in hope of sequestering time, the elusive phenomenon that always seems to escape our grasp.

Many of the most distinguished thinkers in history have wrestled with the concept of time. St. Augustine, a venerable scholar and guiding light for much of Western thought, once pondered the question, "What is time? If no one asks me, I know. If I want to explain it to a questioner, I do not know." Seventeen centuries later, philosopher-scientist Alfred North Whitehead was only able to add his frustration to Augustine's bewilderment: "It is impossible to meditate on time . . . without an overwhelming emotion at the limitations of human intelligence."

How is it that something so basic as time is so little understood and so difficult to explain? Psychologist John Cohen suggests, "We are dealing here with a profound mystery, in the best sense of the word, which lies at the heart of human experience, on the one hand, and in the nature of things, on the other." Yet it is important to probe the hidden dimensions of that mystery, for we cannot fully begin to understand ourselves, our consciousness, and our culture until we have found a way to understand time in all of its various forms. As time defines us, we need to define time.

A battle is brewing over the politics of time. Its outcome could determine the future course of politics around the world in the coming century. The new temporal warfare is a direct outgrowth of another, earlier battle—an economic, social, and political controversy centering around a long-revered spatial metaphor that "Bigger is better." This cardinal concept, which so dominated our thinking after World War II, started to come under attack from many quarters during the 1960s and 1970s. The industrial nations of the West had organized the future with bigness in mind. Centralization, concentration, and accumulation became the watchwords of modernity. Everywhere there was a mad scramble to develop and enlarge. Conglomeration became passion and mission, absorbing the driving energies of the industrialized age. The spatial bias was best expressed in the often-used label, "Under one roof."

Building up, expanding out, filling in—a world of gigantism had cast a menacing shadow over earthly affairs. Bigness went largely unchallenged, allowed to roam at will across the cultural terrain, both public and private, until at long last small bands of Davids began to appear at random spots along the political landscape, firing off their individual salvos, hoping to redirect the thinking of the age.

Big government was attacked for being wasteful and uncaring. Big business was attacked for being impersonal and greedy. Big cities were attacked for being anonymous and insensitive. A spatial heresy began to take hold, winning over legions of converts to a new vision. "Small is beautiful" began to challenge the once-powerful myth that bigger is better. New spatial metaphors entered the political vocabulary as people began to call for a decentralization of government, a democratization of business, and a diffusion of population. Spatial politics dominated the political landscape and the question of scale divided people along the political spectrum. The appropriate scale of government, business, and community, of science, technology, and military preparedness became the object of intense debate.

Yet even as the spatial battle continues to rage on, unresolved, an equally profound temporal battle is now beginning to take hold. If centralization, concentration, and accumulation epitomized the bigger-is-better theme of spatial politics, then efficiency and speed characterize the time values of the modern age. For a long time, the notion of efficiency and speed enjoyed the same kind of unqualified enthusiasm as the notion of gigantism. If bigger was better, then faster and more efficient was more effective.

The idea of saving and compressing time has been stamped

into the psyche of Western civilization and now much of the world. Time, like space, is perceived as a premium, a rare resource that is used to shape and mold the social life of the nation in ever more sophisticated ways. Modern man has come to view time as a tool to enhance and advance the collective well-being of the culture. "Time is money" best expresses the temporal spirit of the age.

As society at large careens toward the high-speed culture of the twenty-first century, small pockets of protest have begun to appear at scattered outposts along the way to wage battle against the accelerated time frame of the modern age. The new time rebels advocate a radically different approach to temporality. These heretics are challenging the notion that increased efficiency and speed offer the best time values to advance the well-being of the species. They argue that the artificial time worlds we have created only increase our separation from the rhythms of nature. They would ask us to give up our preoccupation with accelerating time and begin the process of reintegrating ourselves back into the periodicities that make up the many physiological time worlds of the earth organism.

Advocates of the new time politics eschew the notion of exerting power over time. Their interest is in redirecting the human consciousness toward a more empathetic union with the rhythms of nature. They believe that if we are to "resacralize" life, we must first "resacralize" time. It is by revaluing the time of each other and by understanding and accepting the inherent pace, tempo, and duration of the natural world that we can offer our species the best hope for the future.

Already a multitude of new movements and constituencies are emerging, each incorporating elements of a new radical temporality. The environmental movement, the holistic health movement, the biological agriculture movement, the animal-rights movement, the appropriate technology movement, the Judeo-Christian stewardship movement, the eco-feminist movement, the bioregionalism movement, the economic democracy movement, the alternative education movement, the disarmament movement, and the self-sufficiency movement are all groping toward a new temporal vision.

While the new time rebels acknowledge that increased efficiency has resulted in short-term material benefits, they argue that the long-term psychic and environmental damage has outstripped whatever temporary gains might have been wrought by the fanatic obsession with speed at all costs. These time heretics argue that the pace of production and consumption should not grossly exceed nature's ability to recycle wastes and renew basic resources. They argue that the tempo of social and economic life should be made more compatible with nature's own time frame. In the years to come, these new time heretics will become a political force to reckon with as time becomes a central political battleground around the nation and the world. Politics, long viewed as a spatial science, is now also about to be considered as a temporal art. The politics of territory is about to be joined by the politics of temporality.

So much attention has been given to the spatiality of politics over the centuries that the temporal aspects have been all but ignored. This work is intended to redress the imbalance by focusing on the unexplored temporal dimensions of the political process. A better understanding of the politics of time can provide a much-needed foundation for the eventual synthesis of space/time politics.

As we explore the various temporal dimensions of reality we will come to understand how time values have played a critical role in precipitating the environmental, economic, social, and spiritual crises that now threaten the very existence of the world community. If we are to extricate our generation and free our children's generation from the spectre of dissolution, we need to develop a far more sophisticated understanding of the politics of time.

In Part One, we will establish the temporal context for the coming battle over time values. We will begin by examining the accelerated nanosecond culture brought on by the introduction of computer technologies. We will then contrast the artificial rhythms of our high-speed culture to the organic rhythms of nature in a survey of the new field of chronobiology. Finally, we will compare the fast-paced computer culture with traditional cultures that are more attuned to the biological and physical rhythms of the planet.

In Part Two we will see how Western civilization has increasingly separated itself from the rhythms of nature with the adoption of a series of novel time-allocating devices. Biotic rituals, astronomical calendars, clock schedules, and now computer programs have all been used to cordon off social time from natural time and to bind the human community to the dictates of those on top of the social ladder. We will examine how this succession of time tools paved the way for the emergence of efficiency as the dominant time value of the modern age.

In Part Three, we will observe how those in power convince the people to accept the time restraints imposed on them by offering them the assurance of a future reward commensurate with the sacrifices being made. The people are told that in return for sacrificing their time, they will be assured access in the near or distant future to an idyllic "timeless" kingdom. Most states construct a paradisiacal image of the future for the people to rally around and strive for.

In Part Four, we will see how those in power legitimize the way they manipulate and regulate social time by contending that a similar process drives the natural order. Leaders argue that the way they organize the temporal affairs of society is reflective of the way the universe itself is organized.

In the last section we will challenge the prevailing temporal orthodoxy with a call for the democratization of time. In the process, we will confront the "timeless" illusions and cosmological reductionism that have allowed those in power to maintain their hold over the time dynamics of society. In looking toward the future, we will critically challenge the hyperefficient nanosecond time world that is emerging and advocate a radically different approach to the organization of social time that is more congenial to the time orientation of the natural world. Finally, we will examine the shift in the political spectrum away from the traditional spatial metaphors of right and left wing to a new temporal spectrum with empathetic rhythms on one pole and power rhythms on the other. Those aligning themselves with the empathetic time frame are committed to the development of an economic and technological infrastructure that is compatible with the natural production and recycling rhythms of the earth's ecosystems. Advocates of the power time frame favor a high-technology vision of the future where the rhythms of nature are totally subsumed by the accelerated rhythms of a fully simulated environment.

Time wars will increasingly dominate the politics of tomorrow. It is important, then, to be able to reconceive the art of politics, economics, and culture in temporal terms in order to effectively wrestle with the many new themes that will present themselves in the years ahead as the human race struggles to redefine its relationship to time.

J. T. Fraser has said that "the *Weltanschauung* of an individual and of an age, that is the perception of life and concept of things preferred, is essentially a view of time." To change ourselves and usher in a new order of the ages, we must first be willing to redefine our thinking about the nature of time and our personal and political relationship to it.

PART I

THE
TEMPORAL
CONTEXT

1

THE NEW
NANOSECOND CULTURE

It is ironic that in a culture so committed to saving time we feel increasingly deprived of the very thing we value. The modern world of streamlined transportation, instantaneous communication, and time-saving technologies was supposed to free us from the dictates of the clock and provide us with increased leisure. Instead there seems never to be enough time. What time we do have is chopped up into tiny segments, each filled in with prior commitments and plans. Our tomorrows are spoken for, booked up in advance. We rarely have a moment to spare. Tangential or discretionary time, once a mainstay, an amenity of life, is now a luxury.

Despite our alleged efficiency, as compared to almost every other period in history, we seem to have less time for ourselves and far less time for each other. Even the idea of savoring an experience has become an anachronism in a world where "being" is less important than "becoming" and where expedience is a substitute for participation.

Clearly we have had to pay a heavy price for our efficient society. We have quickened the pace of life only to become less patient. We have become more organized but less spontaneous, less joyful. We are better prepared to act on the future but less able to enjoy the present and reflect on the past. We have learned how to extract and make things at a faster pace but only end up exploiting and devaluing each other's time at the workplace in order to increase production quotas. The efficient society has increased our superficial creature comforts but forced us to become more detached, self-absorbed, and manipulative in relation to others.

The desire, especially of the Western world, to produce and consume at a frantic pace has led to a depletion of our natural endowment and the pollution of our biosphere. Nature's own production and recycling rhythms have been so utterly taxed by the twin dictates of economic efficiency and speed that the planetary ecosystems are no longer capable of renewing resources as fast as they are being depleted, or recycling waste as fast as we discard it.

Statistics tell the grim story of a civilization hell-bent on saving time on the one hand while eliminating the future on the other. Most of us have dismissed the warnings of doomsdayers because of an inherent optimism or unwillingness to change our profligate ways. Nonetheless, the proliferation of nuclear bombs, the mass extinction of plant and animal species, poisoned water, fouled air, and eroded land serve as a constant and dangerous reminder of the toll that has been exacted from the future in the name of today's progress.

As the tempo of modern life has continued to accelerate, we have come to feel increasingly out of touch with the biological rhythms of the planet, unable to experience a close connection with the natural environment. The human time world is no longer joined to the incoming and outgoing tides, the rising and setting sun, and the changing seasons. Instead, humanity has created an artificial time environment punctuated by mechanical contrivances and electronic impulses: a time plane that is quantitative, fast-paced, efficient, and predictable.

The modern age has been characterized by a Promethean spirit, a restless energy that preys on speed records and shortcuts, unmindful of the past, uncaring of the future, existing only for the moment and the quick fix. The earthly rhythms that characterized a more pastoral way of life have been shunted aside to make room for the fast track of an urbanized existence. Lost in a sea of perpetual technological transition, modern man and woman find themselves increasingly alienated from the ecological choreography of the planet.

Today we have surrounded ourselves with time-saving technological gadgetry, only to be overwhelmed by plans that cannot be carried out, appointments that cannot be honored, schedules that cannot be fulfilled, and deadlines that cannot be met.

Strangely enough, even as society finds itself incapable of catching up with the time demands of the modern age, a new and faster time technology is being introduced into the popular culture—a technology that threatens to accelerate our sense of time beyond anything we experienced during the short reign of the modern age.

It is likely that within the next half-century, the computer will help facilitate a revolutionary change in time orientation, just as clocks did several hundred years ago when they began the process of replacing nonautomated timepieces as society's key time-ordering tools.

The new computer technology is already changing the way we conceptualize time and, in the process, is changing the way we think about ourselves and the world around us. We are entering a new time zone radically different from anything we have experienced in the past. So different is the new computer time technology that it is creating the context for the emergence of a new language of the mind and an altered state of consciousness, just as the automated clock did in the thirteenth century when it laid open the door to the Age of Mechanism and the spectre of a clockwork universe.

If it is difficult to think of the computer as introducing a new time orientation of equal historical importance to the introduction of the clock, perhaps it is because the futurists, business leaders, and technologists have been defining the new tool in purely material terms. It is not uncommon to hear the computer being compared to the introduction of the steam engine. The steam engine replaced muscle power with an inanimate form of energy, giving rise to industrial production. In a like vein, it is argued that the computer complements the human mind with the aid of an artificial intelligence, giving rise to what futurists call the postindustrial or information age. In the rush to extol the economic virtues of the new computer technology, futurists have failed to recognize a deeper, more important purpose embedded into the operating principles of humanity's newest artifice. The importance of the computer lies well beyond the vast extension of material benefits it promises to produce. Underneath all of the material projections lurks a new temporal projection, and it is here in this temporal realm that the long-term impact of the computer will be most acutely felt by civilization.

In just two decades, the computer has invaded every aspect of our culture and transformed our way of life. It is estimated that by 1990, nearly 50 percent of all American workers will be using electronic terminal equipment. In addition, some thirty-eight million terminal-based work stations will be on-line in offices, factories, and schools. Nearly thirty-four million households are expected to have home computers by the next decade, while another seven million portable terminals will be in use.[1] Computers are fast becoming staples, finding their way into every conceivable nook and cranny of modern life. They are changing the way we work, play, communicate, and socialize. They are changing our environment and our relationship to it. And most important, they are changing our relationship to time.

The computer introduces a new time perspective and with it a new vision of the future. We are so used to telling time by means of the clock that our minds rebel at the prospect of adopting an entirely different form of timekeeping. While at this emergent state it is difficult to grasp, or even to imagine, the full impact of a shift in time reckoning from clock to computer, an examination of the distinguishing features of this new timepiece provides a clue to the changes in time consciousness that lie ahead.

To begin with, the clock measures time in relationship to human perceptibility. It is possible to experience an hour, a minute, a second, even a tenth of a second. The computer, however, works in a time frame in which the nanosecond is the primary temporal measurement. The nanosecond is a billionth of a second, and though it is possible to conceive theoretically of a nanosecond, and even to manipulate time at that speed of duration, *it is not possible to experience it*. This marks a radical turning point in the way human beings relate to time. Never before has time been organized at a speed beyond the realm of consciousness.

The author Tracy Kidder describes how one computer engineer relates to nanoseconds:

> I feel very comfortable talking in nanoseconds. I sit at one of these analyzers and nanoseconds are wide. I mean you can see them go by. Jesus, I say, that signal takes twelve nanoseconds to get from there to there. Those are real big things to me when I'm building a computer. Yet, when I think about it, how much longer it takes to snap your fingers, I've lost track of what a nanosecond really means. (The snap of a finger is equivalent to the passage of 500 million nanoseconds.)[2]

As we move into fifth- and sixth-generation computers, this new concept of time will pose a variety of problems. Computers of the twenty-first century are likely to be able to make decisions on a wide range of activities in nanosecond time. When many of the decision-making activities of society take place below the threshold of human consciousness, social time, as measured by the clock, becomes irrelevant. The events being processed in the computer world exist in a time realm that we will never be able to experience. The new "computime" represents the final abstraction of time and its complete separation from human experience and the rhythms of nature.

Many people first experience the difference between the worlds of computime and clock time when playing video games. Sociologist Sherry Turkle interviewed video game players and found that the sense of time urgency and intensification were often cited by users as compelling features of the computer-programmed games.

"The game is relentless in its demand that all other time stop and in its demand that the player take full responsibility for every act. . . ."[3] Computer games draw the user into the exclusive time frame established by the program. The pace is set by the game. As Turkle points out, "The rhythm of the game belongs to the machine, the program decides."[4] Unlike pinball games, or even more static games like Monopoly, in which the player can affect or at times even dominate the various time elements, in computer games the user must surrender totally to the tempo of the program. As one twelve-year-old remarked in an interview with Turkle, you have to play to "the heartbeat" of the computer if you are to be successful at the game.[5] The ultimate goal in most computer games is to secure more time. Every aspiring video game player dreams of the perfect game, the game that goes on forever.[6] Time, in computer video games, is a foil, a resource, and a prize all wrapped up as one.

The really good video game players are able to block out both clock time and their own subjective time and descend completely into the time world of the game. It is a common experience for video game junkies to spend hours on end in front of the console without any sense whatsoever of the passage of clock time. According to Craig Brod, one of the growing number of psychologists specializing in computer-related distress, "those who live with computer workers invariably complain that disputes over time are a major source of friction."[7]

Long-term computer users often suffer from the constant jolt back and forth between two time worlds. As they become more enmeshed in the new time world of the computer, they become less and less able to readjust to the temporal norms and standards of the traditional clock culture. They become victims of a new form of temporal schizophrenia, caught between two distinctly different temporal orientations.

Psychologists and sociologists have begun to study the effects of the new computer time conception on both individual personalities and the society at large. Their findings, though still preliminary, suggest that the battle over changing time orientations may well become a central social concern of the next century.

In his recent book, appropriately entitled *Silicon Shock*, information specialist Geoff Simons draws an interesting analogy that captures the awesome speed of computer time:

> Imagine . . . two computers conversing with each other over a period. They are then asked by a human being what they are talking about, and in the time he takes to pose the question, the two computers have exchanged more words than the sum total of all the words exchanged by human beings since *Homo sapiens* first appeared on earth 2 or 3 million years ago.[8]

According to Craig Brod and other psychologists, many people have so accommodated themselves to the new sped-up time frame of the computer that they have become impatient with the slower durations they must contend with in the everyday clock culture. In clinical case studies, psychologists have observed that computer compulsives are much more intolerant of behavior that is at all ambiguous, digressive, or tangential. In their interaction with spouses, family, and acquaintances, they are often terse, preferring simple yes-no responses. They are impatient with open-ended conversations and are uncomfortable with individuals who are reflective or meditative. Computer compulsives demand brevity and view social discourse in instrumental terms, interacting with others only as a means of collecting and exchanging useful information. Above all, they put a high premium on efficient communication. For that reason, says Brod, "They prefer to communicate with people who are 'system literate' so as to transfer information quickly. People who talk too slowly or in general terms are avoided or ignored."[9]

In work situations, computer compulsives find it exceedingly difficult to switch back and forth between the computer time world and the slower clock time world. In the service industries, quality control experts are beginning to observe the effects of the speedup on the interactions between sales personnel and customers. The former complain that the oral responses of customers are too halting and imprecise, that it often takes minutes to secure needed information from customers that could be processed in seconds by the computer. Customers are badgered to be brief and exact so that the information needed to complete the transaction can be processed expeditiously by the computer.

Computer compulsives do not easily tolerate interruptions when they are "interfacing" with the computer. Every encroachment threatens to break their concentration and their transformed time consciousness. In their new time world, they find themselves involved with programs measured in nanoseconds while the outside world is trying to drag them back up to the surface of "reality" and its traditional schedules of minutes and hours. These time pioneers become anxious; their desire is to travel forward into the new time terrain and to resist attempts to force them back into the conventional time frame.

At work, computer compulsives resist the clock culture that relies heavily on meetings and conferences. They are uncomfortable with extended interpersonal interaction and often feel that appointments interrupt "their mission to generate code and work intensely with a program."[10] One programmer at a major software company says that he finds staff meetings and socializing with other workers to be an annoyance and is often angered by interruptions of his work. He explains his feeling by means of a telling analogy. "It's like making love," he says of programming. "If you're making love, you wouldn't want to be interrupted."[11]

Computer time is different from clock time in still another

important respect. The anthropology of time is rich and varied, but in every culture we know of, the temporal ordering of human relationships has always centered on face-to-face interaction. It is true, of course, that over the past four thousand years the human family has developed an increasingly sophisticated array of tools to facilitate communication from a distance. Script, print, and telecommunications have lessened the need for face-to-face interaction. Each of these great revolutions in communications has radically altered our sense of time. Yet in every culture up to now the temporal order has been established primarily around face-to-face interaction, with other forms of communications existing as extensions of that interaction. Now computer technology threatens to change those priorities by incorporating communication and time ordering in one device.

The computer is a form of communication like script, print, and the telephone, but it is also a time tool, like the clock on the wall. As a form of communication, it allows people to engage in a wide range of day-to-day activity without ever coming into intimate contact. With teleconferencing, electronic mail, and office, home, and mobile terminals, there is less need of face-to-face interaction to order the sequence, duration, rhythm, and tempo of modern life. As a timepiece, the computer also establishes a new set of accelerated temporal demands on human behavior. By asserting greater control over the flow of communications between people, and by vastly speeding up the temporal dimensions of that flow, the computer is able to replace face-to-face interactions. In the society of the future, people will increasingly communicate with each other and order the various time dimensions of social and economic life "through" the computer.

Already, the sense of isolation wrought by the introduction of the computer is being felt. As people rely on the computer as both a form of communication and a time frame through which they relate to their fellow human beings, they are finding it more and more difficult to slip back into face-to-face interchange. Psychologists warn that people are losing the traditional temporal skills that have allowed them to relate to each other intimately. The ability to intuit the proper sequences of behavior, knowing how long things should take, being able to align one's own rhythm to that of the group, and being able to synchronize individual and group behavior have become difficult and strained. Psychologists report that computer-compulsive patients feel awkward around people. It is as if they have lost the ability to adjust their behavior to others. Such people have become molded to the temporal dimension of the computer world; they relate to their fellow human beings by and through their new tool.

The computer "nerd" has recently become a familiar archetype in the popular culture. He is portrayed as an individual who speaks too fast, interrupts others, converses in short, choppy sentences, uses non sequiturs, and is out of sync with his physical surroundings. Behind this parody of the compulsive computer savant lies the problem of the new time consciousness. The computer nerd has integrated the time frame of the computer world so thoroughly into his psyche and personality that he is unable to interact effectively with the sequential behavior, durational norms, rhythms, schedules, and coordination patterns of the still-dominant clock culture.

Craig Brod gives an account of the behavior of a top computer expert at a West Coast university, whom he calls Dr. McCarthy, who commonly engages in abrupt exchanges with colleagues:

> One researcher ended a sentence and turned to hear McCarthy's response. But McCarthy had disappeared. Two days later, the researcher was standing near the same spot. McCarthy walked up without greeting and resumed the conversation midthought.[12]

Brod says that McCarthy's sense of sequential and durational behavior is becoming more commonplace among computer people who "relate to others entirely in terms of information exchange," with little regard for traditional codes of civility.[13] The kind of behavior McCarthy exhibited with his coworker is not dissimilar to the way many computer experts interface with their machine.

The following account of both the "wait state" and "time-sharing" provides some insight into the way McCarthy and others handle sequences and durations. In computer jargon the "wait state" refers to the time when the central processor is performing no useful work. It is neither receiving data nor transmitting it, but simply "idling," waiting for a new assignment. Computer experts loathe the very idea of idleness and therefore arrange to have the processor "work on several programs at a time, giving a split-second's exclusive attention to each, depending upon its needs."[14] The technique is called "time-sharing":

> When one program runs temporarily out of data or requires a response from a human operator at a keyboard, the processor does not wait for the potentially slow reply. After all, the ten seconds a human operator may consume in responding represents millions of additions for even a minicomputer. Instead, it goes on to the next program and periodically returns to the first one to see whether there is now more work to be done.[15]

McCarthy utilized the concept of wait time and shared time with his coworker, suspending his "interface" for forty-eight hours to allow his respondent time to process the new information. He then returned to the exact place he had broken off contact to retrieve and process the new information that had been collected in the interim. In this way, time was used in the most economical manner possible, with no time lost in the exchange process.

As people do more and more of their daily business by computer, the amount of face-to-face interaction is radically reduced and the pace of social activity is greatly speeded up. The result, says Geoff Simons in *Silicon Shock*, is that "we are seeing the gradual destruction of human-to-human contact, the elimination of traditional social intercourse, the projection of a new model for human life in which individuals work and play in contact with computer terminals rather than people."[16]

The new time world of the computer is taking its toll on the first generation of computer workers.

A 1981 study by the National Institute of Occupational Safety and Health reports that "clerical workers who use computers suffer higher levels of stress than any other occupational group— including air traffic controllers."[17] Workers weaned on the clock and schedule do not easily adapt to the new time world of computers and programs. Worker resistance has been widespread, but has not reached the explosive level experienced in the early industrial era when an agrarian work force was re-entrained to the temporal rigors of factory clock time. Occasionally a story of computer-induced violence will appear in the news headlines— a worker sabotaging a sophisticated computer program, or destroying millions of dollars of computer equipment. For the most part, however, computer resistance has been more psychological than physical, more passive than active. Still, employers are worried. Much of today's work force is uncomfortable with the new technology and, computer compulsives notwithstanding, reluctant to embrace it with the kind of unbridled enthusiasm its designers had expected.

In an effort to address the problem, employers and software companies have begun to devote considerable time and money to allaying public concern. An avalanche of books has flooded the marketplace, attempting to convince workers and consumers that the new technology is "user friendly." Corporations are providing an increasing number of in-house programs calculated to ease the transition into the new computer world. When all else fails, workers are warned that if they refuse to adjust to the new technology, they will likely be out of a job. Being "computer literate," they are told, is essential to job security and advancement in the Information Society.

Even with all of the carefully planned reeducation seminars, the helpful and inspirational how-to books, the slick advertising campaigns, the well-thought-out marketing strategies, the veiled and not-so-veiled threats of loss of employment, many workers remain anxious, unreceptive, and unconvinced of the computer's blessings. They find it difficult to integrate the computer into their lives or, more accurately, to integrate their lives into the computer.

As was true in the early days of the industrial revolution, employers and public officials are again turning their attention to educating children, realizing that their hopes for an effective work force rest with the next generation. Computer teaching is being introduced into the nation's school system with missionary zeal. In 1980 only a smattering of elementary and secondary schools were equipped with computers and computer curricula. By 1990, virtually every school in the country is expected to be outfitted with the new technology. The entire educational system is being revamped to accommodate this newest technology. The long-term impact on the way children learn and think, say educators, will be truly revolutionary.

The first children of the computer age will soon provide the information society with a willing and eager work force, a labor pool that will have grown up with the computer and understands its tongue as their first language. For these workers of the twenty-first century, the computer will not be an unwanted appendage. Rather, it will be a necessary and vital component of their lives. These workers will have successfully integrated themselves into the time world of computers and programs, so much so that the new time frame will go unquestioned. It will appear as natural for them as the clock and schedule culture is today.

Aware of the great potential of the school system as both a market in which to sell and as a training ground for recruiting the next generation of workers, information technology corporations have invested vast funds into the development of educational software. Jerry Mander of the Public Media Center explains the effect of this development:

> When they [corporations] can . . . provide the computer programs that every kid interacts with, especially in the absence

of human beings to mitigate the process, we will be much closer to a unified field of knowledge, narrower than at present . . . and consistent with corporate values.[18]

Educators are now studying the effect of the computer on the temporal development of children; their findings already suggest that the next generation is beginning to march to a far different beat. First of all, computer children tend to lose the sense of clock time in a new and different way. One twelve-year-old explains his loss of normal temporal awareness by analogy to a dream state. He says that being caught up in the computer is like "falling asleep and thinking you've only been asleep fifteen minutes but you've actually been sleeping the whole night. You try and figure out where the time went. It went into the computer."[19] Craig Brod says that most parents and teachers mistakenly identify the intense involvement with the computer as a psychological "addiction" when, in fact, it is not an addiction at all, but a time warp. According to Brod, "the child has realigned his or her measurement of time" to the time frame of the computer.[20]

When children interface primarily with the computer, sequences, durations, and rhythms speed by, requiring a level of sustained mental concentration that far exceeds what children normally experience when they are learning time skills in conventional settings. The child becomes entrained to the time orientation of his artificial companion, rather than to the more organic time orientation of other children and teachers.

Consider, for example, the LOGO program, perhaps the best-known educational software for young children. With LOGO, a child can program a flock of birds and then put them in motion on the screen. The child follows the motion of the birds, carefully scrutinizing the way they flap their wings, the way they move. But as John Davy observes in his critique of the LOGO program, this is not the same experience the child would gain from watching a flock of birds in nature. On the screen the temporal orientation of the birds is determined by the program. The child fuses with an artificial set of sequences, durations, rhythms, and synchronized activities and patterns. As Davy points out, there are "no smells or tastes, no winds or bird song, no connection with soil, water, sunlight, warmth, no real ecology . . ."[21] All of the environmental cues so essential in the formulation of everyday temporal skills are completely absent.

Harriet Cuffaro offers another illustration of the different sense of temporal entrainment that ensues in computer learning, as opposed to experiential learning in a nonsimulated environment. She uses the example of parking a car. If a child uses blocks as play pieces to park a car, his or her temporal skills will develop quite differently than if the child uses computer symbols. With the blocks, "the child's eye-hand coordination must also contend with the qualitative, with the texture of the surface on which the car is moved, and with the fit between garage opening and car width."[22] Cuffaro points out that "such complexities do not exist on two-dimensional screens."[23] Parking a car on the computer screen is pure action in a vacuum, "motion without context. . . ." Missing from the child-screen interface are the "full-bodied movements characteristic of young children as they interact directly with the environment."[24] It is these full-bodied movements interacting with other phenomena in the life-size real world that have traditionally shaped the temporal skills of children.

In the past, symbolic learning, or abstract learning, has always figured prominently in child rearing. What is different with the emergence of the new computer technology is that the symbolic has now been animated, giving the appearance of reality. Computer learning—with its reliance on simulated nature, electronic birds and flowers rather than the real thing—is beginning to precede experiential learning and is becoming a substitute for it.

Children who have been immersed in the time world of the computer are often unable to readjust to the slower-paced time world of clock culture. Nowhere is this more apparent than when it comes to learning how to reflect, one of the essential time skills taught in the clock culture. Reflection is too slow and confining, too static and dull; it fits the old time world, where memory was important. In the computer world, reflection is as close as the flick of a keystroke. Pondering for long periods of time appears to be uneconomical and unnecessary to a child who has come to think of the past as a code that can be instantaneously called up whenever past information is necessary to fulfill a momentary need.

For this reason, book reading is particularly unappealing to the avid computer child. With reading, the child has to take time to reflect on the story. He has to move into the character and plot and then periodically remove himself in order to think about what has taken place and how that has affected what is currently going on and what is likely to emerge. Book reading requires moments of active engagement woven together with reflective pauses. Computers, however, require constant engagement. The child's mind is never allowed to stray from the immediate action unfolding on the screen. One thirteen-year-old put it this way: "In a computer you're actually there doing it, instead of reading about something that's happening."[25] Of course, this is a strange type of active involvement because the world the child is engaged in is totally contrived, one in which sequences, durations, and rhythms are pure mental constructs removed from the action of the outside world.

Books are not the only component of the traditional school environment falling victim to the new computer time world. Children also complain of teachers being too slow and laborious in comparison to their computer teaching companions. Says one nine-year-old boy:

> Atari cartridges are neat! They come on and tell you what to do. They make it simple. Teachers talk slower than Atari, sometimes they make me angry. I think, "Come on, I want to go back to Atari. It tells me things faster than you do."[26]

As the child moves deeper and deeper into the microworld of electronic circuits and programs, he or she becomes estranged from the time dimension of the natural order, preferring instead to live within the artificial time frame of the silicon chip. Jerry Mander captures the enormity of the loss for future generations: "Nature barely moves at all" in comparison to the computer time world. "It takes an extreme degree of calm to perceive things happening in nature, and I suspect we may be producing a generation of people too sped-up to attune themselves to slower natural rhythms."[27]

This greatly accelerated time orientation is going to profoundly affect every aspect of our culture in the coming century. For tens of thousands of years the human species maintained an organic bond with the pulse of the natural world. While even the earliest societies attempted to impose a social sense of time on top of the biotic and astronomical time frames that order the world and the universe, the human family has never strayed very far from the periodicities of nature or cosmos.

Until the modern era, every concept of time acknowledged an intimate relationship between the rhythms of social life and the rhythms of the earth's ecosystems. Our ancestors relied on the important temporal events in nature, paying close attention to the changing seasons and the changing constellation of stars in the heavens. Human beings marked time by reference to natural phenomena: the time of the rooster crowing, the time of the passing sun, the time of the phases of the moon, the time of the incoming and outgoing tides, the time of the snake shedding its skin, the time the sap runs in the trees, the time the bees take to nectar, the time the birds migrate to far-off places, and the time they return.

While our biological life remains set to the unchanging rhythms of the natural world, our social life has become more and more acclimated to the nanosecond time frame of the computer. The ever-widening schism between natural time and social time is setting the stage for a dramatic confrontation over temporal choices and priorities in the years ahead. To better understand the elements of this emerging conflict, it is essential that we turn our

attention to the biology of time. Nature has its own time orientation, a rich labyrinth of rhythms and tempos that unite the
physical and biological worlds into a synchronized temporal weave.
It is impossible to grasp the full extent of our temporal alienation
in the new nanosecond culture without first examining the age-
old biological rhythms that animate the very core of our existence.

2

CHRONOBIOLOGY: THE CLOCKS THAT MAKE US RUN

To rediscover time, it is essential to journey down into the microworld far below material surfaces. As we penetrate each material substrate, our sense of reality is jolted by the mercurial nature of what we assumed was hard physical reality. Organs dissolve into tissues, tissues dissolve into molecules, molecules dissolve into atoms. If we could magnify the tiniest aspect of our physical world sufficiently, we would find that even the atoms dissolve. What we discover at this most elemental level of material reality is not hard material things but oscillating fields and waves of rhythms. Below the material world that we have long accepted as bedrock reality lies another world discovered by twentieth-century physics: a nonmaterial world of pure temporality, a world of vibrating forces, a world of energy pulsations interacting rhythmically in an elaborately choreographed dance that seems to spread out and give order and meaning to the whole of the universe.

It is here in this silent, untouchable realm that we uncover a more profound order, one that has been little explained and little understood by human cognition, but that now increasingly calls out for our attention.

For as long as we care to remember, we have been classifying and reclassifying, ordering and reordering the world as if it were made up solely of spatially bound, material bits of reality.

Now we begin a new journey, a probe of the temporal order that underlies, informs, and gives meaning to the physical dimension.

The idea of time has long been of interest to philosophers but to few others in the intellectual community. Today, time is experiencing a renaissance. It has become a hotly discussed topic among psychologists, anthropologists, and sociologists. Nowhere is it receiving more attention, however, than in biology, where hundreds of scientific papers are published each year under the rubric of a new discipline called *chronobiology*. The biologists are bringing the concept of time down from the lofty philosophical perches where it has long resided as the ultimate object of abstract theoretical musing, and are studying it as an observable phenomenon in the physical and biological worlds. With each new discovery in the field of chronobiology, we come closer to redefining ourselves in temporal as well as material terms. The social implications of this metamorphosis in thinking are likely to be enormous and far-reaching.

Like most great shifts in human consciousness, the beginning of this particular historical sojourn into the study of time was inauspicious. A Swiss doctor, Auguste Forel, used to enjoy breakfasting on his terrace. In 1906, Forel made an observation that was to change the course of scientific history. Every morning, at exactly the same time, bees from a nearby hive would show up to sample the jam on his breakfast table. Even after Forel began to breakfast indoors, he would notice that the bees continued to show up on the terrace like clockwork, precisely at the accustomed time, foraging for the jam. Forel concluded that because the bees came only at the hour each morning that coincided with their initial exposure to the jam, they must have some kind of built-in memory of time.[1]

In the 1930s, German scientists L. Beling and O. Wahl took note of the uncanny time sense of the bees in a series of landmark experiments. The researchers observed that whenever a bee discovered a nectar source, it would return the next day at exactly the same sun hour. Lest we jump to the conclusion that the bee was measuring time by the sun's position, it should be pointed out that, even when placed in a cellar or salt mine without any external references whatsoever, the bee would continue to perform on cue.[2]

About the same period, another German scientist, E. Kleber, discovered an interesting rhythm in certain nectar-producing plants. The nectar secretion occurred only at specific times each day. The bees, in turn, discovered what time these flowering plants were secreting nectar and showed up exactly at the appropriate feeding time.[3] Several decades later, scientists tried unsuccessfully to outsmart the bees by moving them via airplane across time zones in an attempt to throw off the bees' biological clocks. Bees trained to collect nectar between 10 A.M. and noon Paris time were flown to New York. There they continued their nectar collecting in phase with the time in Paris.[4]

The bee studies opened up a long-buried dimension of reality. It was no longer possible to entertain the rather elitist idea that only the human species was capable of memory and anticipation, or making and keeping appointments. In fact, when it comes to keeping appointments, a species like the swallow will put even the most organized human traveler to shame. In a small village southeast of Los Angeles, swallows leave a tiny church mission every year on October 23 to travel thousands of miles to the south for winter and then return every spring precisely on March 19. They have been a day late only twice in the past two hundred years.[5]

As scientists begin to probe the mysterious world of biological clocks, they begin to see how perfectly matched the biological rhythms are with the rhythms, tempos, and periodicities of the larger physical environment. Living things seem to be composed of myriad internal biological clocks entrained to work in precise coordination with the rhythms of the external physical world. Living things time their internal and external functions with the solar day, the lunar month, the seasons, and the annual rotation of the earth around the sun. The daily cycles are known as circadian, which means "about a day," the monthly cycles are known as lunar, and the yearly are referred to as circannual.

The first indication of the existence of endogenous circadian rhythms was reported in 1729 by the French astronomer Jean de Mairan. He knew that plants extended their leaves during the daylight hours and folded them at night. He was quite surprised, however, to find that certain plants he observed in their own environment would continue to open and shut their leaves on cue, at the appropriate time, even when kept in total darkness in an enclosed space.[6]

Further scientific evidence of the circadian nature of biological clocks came in 1935. E. Bünning carried out a series of experiments with the fruit fly, *Drosophila*. Under naturally occurring conditions, the adult fly always emerged from the pupal case close to dawn. Bünning found that even if *Drosophila* was exposed to constant light and temperature over many generations the circadian rhythm would remain unchanged, demonstrating that the biological clock was indeed endogenous and impervious to environmental manipulation.[7]

In another set of tests, Curt P. Richter of Johns Hopkins University tried to break the circadian cycle of rats by injecting them with drugs, shocking them with electricity, freezing them, stopping their heartbeat, blinding them, and even removing whole sections of their brains, all to no avail. The rats continued unabated in their twenty-four-hour activity cycle despite the Herculean roadblocks Richter put in their path.[8]

For years, scientists have been preoccupied with the search to locate the source of biological rhythms. They have removed body part after body part trying to uncover the location of these clocks and, still, the clocks keep ticking. The rhythms seemingly defy all attempts to locate their material structure. Psychologist John E. Orme offered perhaps the best explanation for this apparent failure in an essay entitled "Time, Rhythms, and Behavior":

The twenty-four-hour cyclical process is so basic from an evolutionary point of view that all plant and animal cells possess a basic metabolic circadian rhythm. Thus the rhythm is not a property of any particular organ or biological clock. The whole organism, in a sense, is the clock.[9]

Psychologist Leonard Doob best summed up the importance of circadian rhythms in the biological scheme of things. According to Doob, "The adaptive significance of circadian rhythmicity is that it enables the organism to master the changing conditions in a temporally programmed world—that is, to do the right thing at the right time."[10]

As previously mentioned, not all biological clocks are circadian. Many creatures exhibit tidal, lunar, and circannual clocks as well. C. J. Whitrow has catalogued many such examples in his book, *The Natural Philosophy of Time*. For instance, there is the little green flatworm known as *Convoluta*, which rises to the surface of the sand at high tide and then burrows back under when the sand dries. While that might not sound like much of a feat, researchers have found that the same flatworm will maintain the same exact rhythm when placed in an aquarium, far away from any ocean tide.[11]

Then there is the case of the palolo worm. This particular creature reproduces "only during the neap tides of the last quarter-moon in October and November."[12] The brown alga *Dictyota* lays its eggs "nine days and fifteen to sixteen days after exposure to moonlight."[13] Whitrow points out that many of the creatures that have been observed to follow tidal and lunar rhythms will continue to exhibit the same rhythmic arrangement in the absence of these external cues.

The Rocky Mountain ground squirrel is a particularly good example of circannual rhythms at work. In the summer of 1963, K. C. Fisher and E. T. Pengelley placed a squirrel in a small windowless room, provided it with food and water, and set the temperature at the freezing point. From August to October, the squirrel ate normally and maintained its body temperature at a constant 37 degrees Celsius. In October, the squirrel stopped eating and drinking and began to hibernate just as it would have had it been out of doors in its natural habitat. In April, it came out of hibernation and resumed its normal feeding activity just as it would have under naturally occurring conditions.[14]

Biologists admit that time is a fundamental aspect of all living things. For the most part, however, they are of the opinion that time is something that has been imprinted into material existence. In recent years attention has been focused on the genes in an attempt to locate the timing mechanisms in specific nucleic acid sequences. In this way, biology remains steeped in the traditions of nineteenth-century physical science. Molecular biologists, in particular, are still influenced by Newtonian physics with its conception of fixed bits of solid matter interacting in a timeless spatial setting. For these scientists, all reality is conceived of in material terms. It is no wonder, then, that they expend so much effort trying to locate the material structure of biological clocks. They start with the a priori assumption that time clocks are tucked away somewhere inside matter.

If, however, one chooses to start with an understanding of twentieth-century quantum physics, the search for the elusive biological clocks takes a different turn. Below physical surfaces, underneath hard substances, inside dense cores, we are beginning to perceive a new reality, a domain where pulses, rhythms, and periodicities are the rule, the order, and the reality. Instead of perceiving time as one of the components of matter, we perceive the material world as merely an expression of a more fundamental temporal reality.

Time, then, is more than just a feature of reality; it may well be underlying reality. What we perceive as solid, material forms may be a macroexpression of rhythms, vibrations, pulsations, and fields that give rise to and order all physical phenomena.

This new perception of reality allows us to better understand why living things seem to be so finely entrained with the larger rhythms of the physical world. Every level of biological reality, from organs to molecules, is woven from the same temporal patterns that give order to the material universe.

For years, chronobiologists have been feuding over the question of whether biological clocks are endogenous in nature and inherited, or exogenous and influenced by the changing rhythms of the physical milieu. Although the evidence at this time seems to suggest that most rhythms are inherited but can be slightly manipulated by changing environmental cues, a small group of distinguished researchers continues to give greater credence to the exogenous theory. A leading proponent of the exogenous school is Frank Brown of Northwestern University. His research is interesting beyond the debate at hand because of the wealth of information it provides about the subtle temporal forces that affect the material world we inhabit.

Brown believes that all living things are sensitive to and influenced by the many fields that permeate every stratum of physical reality. These include magnetostatic, electrostatic, and electromagnetic fields. He argues that all creatures act as sensitive cosmic receiving stations, continuously responding to and adjusting to the subtle field changes in their environment.

Working with worms and mud snails, Brown performed a series of fascinating experiments designed to show how living things are sensitive to and affected by the earth's magnetic fields. Worms and snails are known to move in given compass directions at different times of the day, month, and year. He placed a bar magnet under a grid while the worms and snails were moving on top of the grid surface. The creatures reoriented themselves in tandem with the changes in the magnetic field. In another set of experiments, Brown and his colleagues found that the worms and snails were able to respond to the strength and direction of weak changes in gamma fields.[15]

Frank Brown believes that all organisms have a way of making use of atmospheric media to anticipate and respond to a changing environment. This might help explain why fiddler crabs are able to disappear into inland burrows twenty-four hours before a hurricane and why elk begin to huddle under trees thirty-six hours before the onslaught of a major blizzard.[16]

It is not yet known for sure whether living things, through the long process of biological history, have developed biological clocks that parallel the geophysical frequencies of the physical world or whether organisms continually adjust their internal rhythms to corresponding geophysical frequencies. Chances are that living things are made up of both endogenous and exogenous clocks. That is, they inherit a set of imprinted rhythms that parallel the physical rhythms of the environment and are also able to attune their internal cycles to changes in the physical fields around them.

Whichever is the case, we are beginning to suspect that below the material world is a temporal reality pulsating with ordered rhythmic activity, and this ordered activity may well play a role in governing the material realm that we interact in. John E. Orme writes:

> The physical universe is basically rhythmic in nature. The moon revolves around the earth, the earth around the sun, and the solar system itself changes spatial position with time. All these phenomena result in regular rhythmic changes, and the survival of biological species depends on the capacity to follow these rhythms.[17]

The human body far surpasses any artificial timing mechanism that human ingenuity has ever designed. Even the best Swiss timemaker could not possibly presume to model a collection of timepieces to rival the precision, coordination, synchronization, and reliability of the many biochemical clocks that work in tandem within the human body. Relentlessly and meticulously, the body's inner pulsations mesh together in an exacting daily ritual, assuring the perpetuation of the human life force.

The number of tasks that need to be choreographed within the human body is awesome. Blood pressure, heartbeat, body temperature, metabolic rates, hormonal secretions, wake and sleep cycles, are only a few of the systems that need to be timed and coordinated with precision if the human body is to function properly.

Many of our most basic internal processes follow a twenty-four-hour cycle. Some of our biological clocks are attuned to lunar cycles, and still others are set by annual cycles. Millions of people have become familiar with the idea of biological clocks as a result of exposure to jet lag and shift work. In both situations, the body's internal rhythms are jolted out of synchronization by the radical temporal changes to which the body is forced to adjust.

Jet lag has become a very real and persistent problem in the industrialized nations. On any given day, people fly across multiple time zones and cause disruption to their internal biological rhythms. Jet lag symptoms can vary but usually include gastro-intestinal discomfort, a decrease in alertness and attention span, an inability to sleep, and a feeling of general fatigue. Athletes competing in international meets, businessmen attending conventions, and diplomats taking part in political conferences often arrive early at their destinations in order to allow time for their biological clocks to readjust to the changes in time zones.

In their book, *The Clocks That Time Us*, Martin Moore-Ede, Frank Sulzman, and Charles Fuller recount a now-famous story in which biological clock desynchronization changed the course of history. In the 1950s, when Secretary of State John Foster Dulles flew to Egypt to negotiate the Aswan Dam Treaty, the negotiations began almost immediately after his flight touched down. The secretary's biological clocks were severely desynchronized, and, as a result, his diplomatic skills fell short of the sensitive task at hand. The dam project was awarded to the USSR, providing them with their first foothold in the Middle East, and changing the balance of power between East and West for nearly a decade.[18]

Even minor changes in time orientation have been shown to have statistically significant impact on human behavior patterns.

For example, studies have shown that traffic accidents increase significantly the week after the clocks are changed in or out of daylight savings time.[19]

One out of every six American workers is involved in shift work, and industry analysts are beginning to assess its effect on the desynchronization of biological clocks. Shift workers' susceptibility to peptic ulcers and other stomach disorders is two to three times greater than the general population's. Their productivity is also significantly lower. According to a spate of studies, worker error peaks between 3:00 A.M. and 5:00 A.M. A study done on the trucking industry found that the chance of an accident increases by 200 percent at 5:00 A.M.[20]

C. F. Ehret's study of shift work in the nuclear power industry has raised the spectre that the near-catastrophic accident at the Three Mile Island facility might have been related to a desynchronization of the biological clocks of the night shift. The employees in charge of the facility when the accident occurred at four in the morning had been rotating shifts around the clock every week for a month and a half. Ehret points out that the sheer complexity of the monitoring activity in a nuclear power facility requires a level of alertness and coherence that might pose a severe hardship on workers whose internal rhythms were already strained by changes in time shifts.[21]

While jet lag and shift work make us aware of the temporal nature of our biological existence, we have still uncovered only a fraction of the many rhythms that permeate the physiology of the human organism. We know, for example, that regardless of how much liquid we consume at particular times during the day, our urine flow follows a circadian rhythm, declining during the nighttime hours.[22] We know that the kidneys function in tandem with the daily revolution of the earth. We know that potassium excretion peaks between 10:30 A.M. and 2:30 P.M. We know that the liver processes its glycogen reserves according to a dependable circadian rhythm beginning by late afternoon and ending between 3:00 and 6:00 A.M.[23]

Our body temperature also rises and falls in a predictable pattern every twenty-four hours. So, too, does our skin temperature. When asleep, skin temperature is higher on the left side of the body, and, during waking hours, temperature is higher on the right side.[24]

Some researchers believe that a correlation exists between an individual's built-in temperature cycle and whether he or she is a "morning" or "night" person. Lawrence Monroe at the University of Chicago found that individuals whose body temperature rises to normal upon wakening are apt to be very alert in their early waking hours while those whose temperatures do not rise to normal until well into the day are generally more alert and sensitive to their surroundings as evening approaches.[25]

Not surprisingly, researchers also find that performance and achievement levels vary during the day, depending partly on the unique temperature rhythm of each individual. At the National Medical Research Laboratories in Cambridge, England, Robert Wilkinson and Peter Colquhoun found that peak performance on tests correlated with peak body temperature during the day and low performance accompanied the lowest body temperature during the day.[26]

Body temperature can also affect our judgment of time. The first to discover the relationship between body temperature and time perception was Hudson Hoagland of the Worcester Institute of Experimental Biology. Hoagland's discovery came quite by chance on a day he was at home attending to his sick wife, who had a temperature of 104. Mrs. Hoagland asked him to go to the drugstore for her. Upon his return twenty minutes later, she insisted he had been gone for hours. Eager to see if there was any correlation between his wife's temperature and time perception, Hoagland asked her to count to sixty at a speed she thought approximated one number per second. Mrs. Hoagland counted to sixty in much less than one minute. Hoagland followed up on this initial experiment and, on each occasion, his wife would count faster when her temperature rose and slower when it fell. Later studies by other scientists confirmed Hoagland's observations.[27]

In another study, a similar correlation between subnormal temperature and underestimation of time was found. A French researcher, Michel Siffre, spent sixty-three days in a cave hundreds of feet below the Maritime Alps without any access to the correct clock time. At the end of his stay, he estimated that he had spent only thirty-six days in isolation. The doctors monitoring the experiment hypothesized that Siffre's gross underestimation of calendar time might have been due to his lower body temperature while underground. His temperature hovered at 96.8 degrees in a prehibernation state.[28]

The relationship between body temperature and time perception led Gay Gaer Luce to ask whether this might account for why children perceive time as moving slowly while old people perceive time as moving swiftly. Luce discovered that in the progression from childhood to adulthood the metabolic rate slows down, evidence that changing time perception at various stages of the life process might be related to changes in biochemical rhythms within the body.[29]

Researchers at the Institute of Chronobiology at New York Hospital have been studying the relationship between body rhythms and aging and have come to some tentative conclusions that support Luce's hypothesis. According to Daniel Wagner, "There's some indication that sometime—and fairly rapidly—in the mid-fifties, there is a shift in the length of some of the internal rhythms." Dr. Wagner believes that this dramatic resetting of internal clocks might well account for why older people nap more, go to sleep earlier, and wake up earlier.[30]

Studies of the female menstruation cycle have also turned up some interesting data on the way biological clocks synchronize their behavior with one another. Several years ago researchers performed an experiment with women living in a college dormitory. At the beginning of the school term the women's menstruation cycles were logged. Their periods began at scattered times throughout the thirty-day cycle. By the end of the school year, however, many of the women were menstruating in complete synchronization with each other. Independent experiments conducted in various places over time have come up with similar findings.

The growing interest in biological clocks has sparked a host of new concerns within the medical community, not the least of which has been a growing debate over the proper time to administer drugs. A new field known as *chronopharmacology* has emerged virtually overnight. Although still in its early stages, chronopharmacology is uncovering important causal relationships between the time drugs are given or surgery is performed and the biological clock orientation of the patient.

Many scientists have come to believe that the time of day a drug is administered may be just as critical to the patient's well-being as the kind of drug being used. At the University of Arkansas, Lawrence E. Sheving injected three hundred mice with leukemic cells. He then divided the mice into twelve groups and administered chemotherapy to each group at a different time of day. Over half of the mice that received chemotherapy at 5:00 A.M. were cured, whereas only 16 percent of the mice that received the same treatment at 8:00 A.M. survived.[31]

Another set of experiments on various cancer patients at the University of Minnesota provided still more evidence that the time of treatment is critical. William Hrushesky found that ten of twelve patients went into remission as a result of administering the chemotherapy at certain key times during the day. Hrushesky concluded that the results of the study could be attributed "partly to the circadian time of the treatment."[32]

Even our moods are influenced by biochemical clocks. While we have always known that how we feel influences how we behave, it is only lately that scientists have been able to show a relationship between specific emotions and specific biological clocks ticking away inside of us. Joseph Bohlen, of the University of Wisconsin, studied what is known as "arctic hysteria," a condition similar to acute psychosis that afflicts Eskimos. Bohlen and his wife collected urine samples and recorded oral temperatures, blood pressure, and pulse changes among a group of ten Eskimos every two hours. They found that Eskimos demonstrated a peculiar annual rhythm in calcium excretion, putting out eight to ten times as much in the winter as in the summer. Calcium plays a key role in the transmission of messages in the nervous system. This led Bohlen to suspect a correlation between periodic emotional illness among the Eskimos and the annual rhythmic changes in calcium excretion.[33]

Because of biochemical clocks, anger and anxiety can vary considerably during the day. Both emotions require extra doses of adrenaline, and adrenaline, like other biochemical processes, adheres to a changing secretion pattern over a twenty-four-hour cycle. It is likely, then, that the expression of anger and anxiety will be greater or lesser at given periods of the day, irrespective of the object that triggers the emotion, simply because the adrenaline flow necessary to incite the feeling varies in intensity.[34]

Perhaps the most intriguing new discovery tying internal clocks with shifts in mood and behavior was reported in 1985 by the National Institutes of Mental Health. It turns out that spring fever, long romanticized in verse and prose, is related to the chemical changes that take place in specific biological clocks in the human body. According to NIMH scientists, the brain measures the length of each day and then uses that information to regulate the secretion of specific brain hormones that affect mood and behavior. During the shorter days of the year, the brain secretes more melatonin, which brings on depression. As the days lengthen toward early spring, the pineal glands secrete less melatonin, thus alleviating the sense of depression brought on by this particular hormone.[35] By exposing severely depressed patients to massive doses of artificial light over an extended period of time, scientists have been able to shift moods dramatically. The artificial light lowers the secretion of melatonin, and in so doing, lifts the veil of depression hovering over the patient.[36]

Researchers have even begun to suspect that a relationship exists between biological clocks and the statistical fact that most births and deaths occur in the early morning hours, that the peak number of deaths from arteriosclerosis occur in January, and that most suicides occur in May and June.[37] As researchers in the field of chronobiology probe deeper into the unknown dimensions of biological clocks, it becomes increasingly apparent that temporal considerations play an essential role in ordering the entire life process. Below material surfaces, life is animated and structured by an elaborate set of intricately synchronized rhythms that parallel the frequencies of the larger universe. Chronobiology provides a rich new conceptual framework for rethinking the notion of relationships in nature. In the temporal scheme of things, life, earth, and universe are viewed as partners in a tightly synchronized dance in which all of the separate movements pulse in unison to create a single organic whole. The idea of biological clocks and circadian, lunar, and circannual entrainment suggests a radical new interpretation of context as more of a rhythmic bond than merely a spatial setting.

While all living things can be characterized by the biological rhythms they inherit, only human beings impose a social sense of time on top of the biological clocks with which we are born. Since the dawn of Western consciousness, we have lived out our lives in a schizophrenic middle kingdom where biological and physical time clash head-on with our cultural and social time. And with every change in rhythm, tempo, and timing of either temporal order, we are forced to mediate a compromise that will allow us to continue to walk the tightrope that separates these two distinct and irreconcilable temporal worlds. If our existence is a dual one, at once both natural and social, inherited and learned, its basis is to be found in the first great separation, that point where we began the process of expropriating our own time, claiming our independence from the great temporal symphony that orchestrates the other worlds we are fashioned from.

The social clock begins ticking shortly after birth. Researchers are now finding that an infant starts off with a deeply ingrained set of biological rhythms that are then manipulated, rearranged, and refined to cohere with the temporal expectations, standards, and norms of the culture. The social manipulation of time is made easier by the fact that a baby is born with an innate ability to synchronize perfectly its own rhythmic movements with external rhythms.

Over a decade ago, William Condon published the amazing results of a study that helped reshape much of our thinking about the interaction between biological and socializing rhythms in the course of development. Having filmed the interplay between adult speech and infant movement, Condon discovered, by isolating the frames, that the baby moved in complete synchronicity with every auditory change in adult speech. Even the most subtle changes in auditory rhythm were accompanied virtually simultaneously by a corresponding shift in movement by the infant. "The body of the listener dances in rhythm with that of the speaker," said Condon.[38] The baby's innate ability to entrain its own biological rhythms with the socialized auditory rhythms of the adult sheds new light on the parent-infant bond. The orthodox image of two discrete entities exchanging messages was subsumed by the reality of an organic bond in which each participant was fused into a single rhythmic arrangement.[39]

A new view of the baby has emerged from temporal studies, and it is a far cry from John Locke's description of the infant as a *tabula rasa*—a blank slate that is given identity solely through the socialization experience. Researchers now believe that a child is born with an already partially developed set of biological rhythms that are then entrained by the parents' interaction with him or her at various stages of temporal socialization. The parents' own particular temporal orientations are likely to play a significant role in the sense of timing imprinted into the infant during the first few months of life.

Freudian psychology has developed a kind of shorthand for explaining the process of mediation that adjusts internal rhythms

with externally imposed temporal norms. Freud posed a fundamental dichotomy, expressed in terms of two all-embracing metaphors: the pleasure principle and the reality principle.[40]

The unconscious is a timeless realm of pure fantasy where
immortality and omnipotence reign supreme. The baby lives in
this paradisiacal state for only the briefest period of time. Social
schedules soon intervene, at first only sporadically and then more
persistently, forcing the infant to experience the frustration of
having to compromise its omnipotence to the dictates of the social
clock. This is where the reality principle sets in. The baby comes
to the realization that it cannot always have every one of its needs
met immediately. It must wait, even compromise. More devas-
tating still, it must submit to demands from the outside, obeying
the temporal requisites forced on it by the culture. As psychologists Edmund Bergler and Géza Roheim have pointed out, "The
pleasure principle and timelessness are linked together, as are
time and the reality principle."[41] The reality of time demands
impinge on the timeless world of pure unadulterated pleasure
seeking, forcing all individuals to accept a measure of compromise if they are to be weaned into the culture.

Socialization requires some degree of renunciation. If every
child were allowed to act on every one of its impulses, there would
be chaos. Society is built on the renunciation of urges. Freud
believed that no other single factor is as important in laying the
foundation for society. "It is impossible to ignore the extent to
which civilization is built upon renunciation of instinctual grat-
ifications."[42]

All renunciation is tolerable to the extent that it is considered
temporary. A child is willing to repress an immediate urge if
reasonably convinced that he can anticipate a comparable reward
sometime later on. Renunciation and anticipation are not unique
characteristics of *Homo sapiens*. Other animals can be entrained
to repress urges and anticipate future rewards as Pavlov's famous
dog studies have clearly demonstrated. The difference, rather, is
one of degree, not of kind. *Homo sapiens* is able to anticipate

future states far better than any of the other animals, and it is this superior anticipatory power that allows our species to renounce more in the short run in return for greater rewards in the long run.

The imposition of culturally derived temporal norms is the key socializing tool. Every culture inculcates its newest members by way of an elaborate and often complex process of temporal entrainment.

Studies over the years have established a fairly accurate picture of the ages at which children pass through critical temporal watersheds in their social development. Between one-and-a-half years to two years, the child lives, for the most part, in the present. Between two and three years of age, the child begins to develop a "temporal perspective" by using words that refer to the past, present, and future. Interestingly, future-oriented words always precede past-oriented words in the child's temporal development. By age four, the child is much more attuned to the future and is able to project himself into future situations. Still, the child's concept of past, present, and future remains disjointed, punctuated by moments of clear temporal lucidity and periods of sheer temporal confusion. Psychologists Lawrence Stone and Joseph Church have observed:

> For all the rapid development of ideas of past, present, and future, of short and long time intervals, even at the end of the preschool period there is still no overall, consistent framework of time, but a patchwork of uncoordinated time concepts.[43]

By the time the child begins school, he or she understands the days of the week. At six the meaning of the four changing seasons of the year is understood. At age seven the child begins to understand the concept of months and the notion of clock time. Between eight and thirteen, the child continues to expand his temporal perspective into both the past and the future.[44]

Between the late adolescent stage and the coming of puberty, most children begin to develop an increasingly realistic approach to the future. They come to see themselves as planners of their own destinies. The future, long regarded as a fantasy realm in which to project wishful thinking, becomes transformed into a place they see themselves occupying as an adult. This particular temporal passage is crossed generally at that stage when children discontinue their interest in "fantasy" occupations like movie star, cowboy, or circus performer and begin thinking and talking seriously about being engineers, doctors, and lawyers. In a study of twenty-five thousand boys and girls, the onset of puberty was most often the critical turning point where childhood career fantasies were abandoned and serious attention began to be paid to realistic occupational roles. For girls the transition occurred at ten-and-a-half and for boys at eleven-and-a-half.[45]

The human species is unique, then, in that it creates its own time. As we have just seen, we begin imprinting the time values of our culture onto our young shortly after their birth. As our children grow up, they are entrained to live in two time worlds simultaneously: the biological time world they inherit and the social time world they learn. Until recently these two time worlds were closely attuned. The accelerated time frame of the modern age, however, is beginning to drive a permanent wedge between the rhythms of culture and the rhythms of nature, threatening a complete break in temporal bonds between the two worlds. To appreciate exactly how far the nanosecond culture has strayed from the periodicities of nature, it is essential to contrast the fast-paced computer age we live in with more traditional cultures where the rhythms of daily life more closely mirror the time frame of the natural world. We, therefore, turn our attention next to the anthropology of time in an effort to better understand the gravity of the temporal crisis at hand.

3

ANTHROPOLOGICAL
TIME ZONES

Historian Daniel J. Boorstin says that when it comes to chronicling human accomplishments to date, we must begin by acknowledging that "the first grand discovery was time."[1] Time has been our most important innovation. We have used this instrument to fashion our cultures. It is the primary socializing tool.

Every culture is, to a great extent, a reflection of the kind of temporal orientation it adopts. No two cultures think exactly the same because no two cultures share identical conceptions of time. Cultures, like individuals, are time-bound and, like the individuals that comprise them, they each take on a unique personality, depending on the particular temporal consciousness they have fashioned.

Scholars have identified six distinct temporal dimensions that are continually at play as individuals interact with each other in a social context. Every thought, event, occurrence, or situation is definable in terms of sequential structure, duration, planning, rate of recurrence, synchronization, and temporal perspective. All of the six primary temporal dimensions exist in every culture. The way a society chooses to define and use each of these building blocks of time determines the overall temporal orientation of the culture. It is important, then, to explore each category of time in depth to better understand the ways in which the time dimension affects the individual and the society.

The best place to start is with the idea of sequence. Every society establishes a temporal standard for most events, an order of regularity for how things should unfold. Sequences deal with what comes before and what comes after. A proper understanding of sequential behavior is critical to effective socialization.

The childhood experience is sprinkled with sequential reminders. Parents are forever admonishing children for failing to heed the proper sequential cues. What adolescent has not heard a parent say, "You must first eat your dinner before you can have dessert," or "You must first do your homework before you can go out and play."[2] In order to interact successfully, we must know what is expected of us at any given time. Much of the learning process is designed to teach us the proper order in which things should be done.

In Western culture, events follow one another in a linear, causal order. We place a premium on enclosure. That is, once a task or activity has been initiated, we feel compelled to see it through to its completion before going on to something else. We do not easily tolerate loose ends and are uncomfortable with the idea of suspending an activity or event in limbo for long periods.

Our ideas about the proper order in which things should unfold differ from some other cultures'. Anthropologist Edward T. Hall recalls a confrontation between a group of Pueblo Indians in New Mexico and state authorities, which highlights the potential problems that can occur when two groups entertain very different notions regarding the proper sequence in which events should unfold. The Indians informed the State of New Mexico that they would close down a state road that went through their land unless the government made suitable arrangements to compensate them for public access and right-of-way. Several years went by without further discussion between state authorities and the Indians. Then, without any warning, the Indians struck. One morning they erected a steel guardrail across the road and attached a sign saying that they were exercising their right to close the road.

The New Mexico authorities were dumbfounded. They could not understand why, after so many years of silence, the Indians had decided to suddenly act on their demands. A spokesman for the Pueblo Indians responded by saying: "I don't know why they were surprised. After all, those signs saying we were closing the road were stacked up against my house for a year and everybody saw them. What did they think those signs meant?"[3]

In the white culture a demand is almost always followed in short order by a confrontation, a crisis, action, and resolution. The Pueblos' temporal orientation is quite different. They found it perfectly appropriate to suspend the sequence, engage in other activities, then return years later to complete what they had begun.

It is not enough, however, to know the order in which things should unfold. We also need to know how long things are supposed to last if we are to integrate our activity with others. Many of the most often-heard questions of childhood deal with duration. "When are we going to get there?"; "How long is this going to last?"; "When will it begin?" and "When will it be over?"; these are the ways the child learns about durations.

Society establishes durational guidelines for a wide range of activities, occurrences, and events, including those of a very personal and intimate nature. Sociologist Lois Pratt, writing in the *American Sociological Review*, provided a graphic illustration of how acceptable durations become institutionalized in the culture. Pratt examined the influence of business norms and practices on the changes in bereavement behavior. She found that the amount of time given over to bereavement for a deceased relative or friend has become increasingly determined by the bereavement policies established by American business enterprises.

Over 90 percent of American companies now grant official time off for bereavement. Most business enterprises, however, have set rigid standards governing how much time the employee can officially grieve before returning to work. Most companies have established three days as the formal bereavement period. Employees are expected to complete their public bereavement within this seventy-two-hour period and return to business as usual at the end of the allotted time.[4]

Companies even prescribe the exact temporal boundaries that

are to be honored during the bereavement duration. Paid leave is not to begin until the death occurs. Employees are not allowed to begin leave during the terminal phase of a relative's or friend's illness. The American Management Association has even drafted guidelines to cover weekend deaths. According to the association, "It is usually expected that when death occurs on a Saturday the employee should return to work on a Tuesday following the normal time for the funeral."[5]

It is interesting to note that the acceptable duration for mourning has shrunk significantly in the past century. In 1927, Emily Post reported that the formal mourning period for a widow was three years. By 1950, the acceptable period of bereavement had plummeted to a mere six months. By 1972, Amy Vanderbilt was advising the bereaved that they should "pursue, or try to pursue, a usual social course within a week or so after a funeral."[6]

The corporate emphasis on speedy bereavement has likely contributed to dramatic changes in Americans' funerary practices. Funeral directors report an increasing trend away from protracted, elaborately choreographed burial rituals. Funeral directors say that shortened services do not make enough allowance for viewing the body or for visitation with the deceased. Whereas processions to the cemetery used to be standard funeral fare, often now a single service is completed at the sanctuary and the body is taken to be buried, without a follow-up ceremony at the gravesite.

Disposal by cremation rather than by burial has also become increasingly popular, as it shortens the amount of time necessary to complete the bereavement ritual. Earth burial represents an ongoing commitment by the living to maintain the gravesite, and thus prolongs the period of bereavement, whereas cremation helps "facilitate abrupt severance of ties with the dead and immediate reintegration of the bereaved into the workplace."[7]

According to Lois Pratt, the data indicates:

> The timing and direction of the trends in funerary ritual are consistent with the timing and direction of business bereave-

ment provisions, thus making it plausible to infer an interplay between business policies and the changes in funeral ritual.[8]

Corporate bereavement policies have even begun to influence bereavement priorities within the family. Most companies restrict bereavement leave to the immediate family—including spouse, children, and parents. Grandparents, grandchildren, brothers- and sisters-in-law, and aunts and uncles are all generally ex- cluded, as are close friends. By limiting bereavement to only a few close family relatives, American business has helped facilitate a trend toward less and less community participation in mourning duties and rituals. Funeral directors and clergy both report a decrease in overall attendance at funeral proceedings over the past several decades.[9]

As the bereavement ritual illustrates, an appropriate lapse of time is affixed to almost everything we do within a social context, from eating and sleeping to working and playing. While some deviation from acceptable norms is tolerable, we will often apply sanctions when someone among us spends too little or too much time with a particular activity. Knowing how long to spend on a given pursuit can make the difference between being able to cope with, or being swallowed up by, the many time demands made on each of us by society.

Cultures differ markedly in the way they establish durations for various activities. Our Western concept of time, which is abstract, external, linear, and quantitative, makes little sense to members of other cultures where durations are measured not by the ticking of the clock, but by the unfolding of environmental events or the ordering of sacred rituals. As one scholar aptly put it, in many non-Western cultures they "don't tell you what time it is; they tell you what kind of time it is."[10]

In a study of African schoolchildren, P. M. Bell found that his pupils were simply unable to measure durations in terms of stan- dardized units of abstract time. When asked how long a two-hour bus journey had taken, some students thought it had taken ten minutes, others thought the journey had taken as long as five or six hours. Although in other testing areas these children were found to be very intelligent, they had no awareness of clock time as a way of measuring the duration of an event.[11]

In many traditional cultures, duration is measured by references to specific tasks rather than by abstract numbers. For example, in Madagascar, when someone asks how long something takes, they might be told that it takes the same time as "rice cooking" (about half an hour) or the time it takes to "fry a locust" (a moment). The Cross River natives of West Africa, when asked how long it took for a man to die, would say, "The man died in less time than the time in which maize is not yet completely roasted (less than fifteen minutes)."[12]

The Nuer in Africa divide the day into specific durations according to the order and time for each task to be undertaken and completed. E. E. Evans-Pritchard reports:

> The daily timepiece is the cattle clock, the round of pastoral tasks, and the time of day and the passage of time through a day are to a Nuer primarily the succession of these tasks and their relation to one another.[13]

Economic considerations play a key role in the establishing of durational boundaries. For example, while we take for granted the notion of a seven-day week, it is not uncommon for a week to vary anywhere from three to ten days, depending on the culture. The week came into existence as a means of establishing a duration between market days. The length of the week varies in direct correlation with the communal needs and realities governing the exchange of goods. Hutton Webster, author of *Rest Days*, points out that:

> The shorter intervals of three, four and five days reflect the simple economy of primitive life, since the market must recur with sufficient frequency to permit neighboring communities,

who keep on hand no large stocks of food and other necessities, to obtain them from one another.[14]

Throughout history, cultures have varied significantly in the amount of time they afford to the same events and activities. Psychology professor Robert Levine came up against the problems that arise when two cultures place different values on the duration of a common experience when he accepted an appointment at a university in Niteror, Brazil, a midsize city located across the bay from Rio de Janeiro. His first day of class was scheduled from ten to twelve in the morning. The class began promptly at ten, at least for Dr. Levine. His students, however, strolled in at various intervals from shortly after ten to shortly after eleven. None of the late arrivers felt particularly uncomfortable about not arriving on time. On the contrary, they appeared relaxed and unhurried.[15]

Equally surprising to Dr. Levine was what happened at the end of the scheduled two-hour session. Only a few students left at noon. The rest lingered on, asking questions or listening attentively. Such behavior is virtually unheard of in an American classroom. Finally Dr. Levine had to take his leave, sensing that several of the students would have stayed on for hours if he did not put an end to the session.

Unlike American students who are rigidly entrained to the dictates of the clock, Brazilian students are much more "laid back," and much less compulsive when it comes to conforming to predetermined durational boundaries. For the Brazilian students, the preestablished time duration of ten to noon is more of a general reference point to coordinate activity rather than an inflexible imperative to be unflinchingly obeyed. When surveyed, Brazilian students defined lateness as 33.5 minutes after the scheduled commencement of class whereas their American counterparts at a college in Fresno, California, said that 19 minutes would be considered late to class.[16]

Interestingly, the Brazilian students felt no anguish over being late. In follow-up interviews, they expressed a certain amount of pride in their late-to-come, late-to-leave orientation. In their culture a value is placed on maintaining a certain open-ended approach to durational segments.

Knowing the order in which things should be done and how long they should take are insufficient temporal criteria to assure the proper functioning of social relationships. We also need to learn how to plan out our activities and establish predictable rhythmic patterns.

Our culture constantly sets aside segments of the future to be used for specific activities. The more complex the social environment, the greater the need to reserve pieces of the future ahead of time. Sociologist Eviatar Zerubavel, of Columbia University, points out that planning is indispensable to social organization:

> If all of us were to behave spontaneously, we would probably not have any form of social organization. Social life requires some coordination among individuals. No social event could ever take place if every individual were to have a say in deciding when it ought to begin. . . . Most social enterprises would have been impossible were it not for the durational rigidification of tasks in accordance with some deadlines.[17]

As in the case with sequencing and duration, planning provides a means of insuring predictability. Communities live by routines and habits. Society cannot function without a degree of certainty built into its temporal fabric. Planning helps provide that certainty.

Where once we planned our social and economic activities to accommodate the changes in the biotic and physical environment we live in, we now plan our activities to correspond with such purely social conventions as fiscal years, eight-hour workdays, two-week paid vacations, academic semesters, and tax day. Our social schedules have become increasingly divorced from the periodicities of the natural world, reflecting our repeated attempts to claim some measure of economic independence from the production and recycling schedules that animate the ecological systems here on earth.

The first thing we do when we decide to plan something is to look at our appointment book or the calendar. When the inhabitants of the Andaman jungle decide to plan something, they first smell the odors in their surrounding environment. The Andamanese have developed a complex annual calendar based on the succession of dominant smells of flowers and trees. They use these reference points to plan the corresponding activity for the appropriate time of the year.[18]

The industrial age is unique in that it maintains only a marginal relationship to the periodicities of nature. Every other culture in history has been inseparably bound to the biological and physical rhythms of the larger environment. For example, in Labrador, like most preindustrial cultures, the inhabitants plan their activities to fit the natural timetable they rely on for sustenance:

> Latter part of June to the end of July: cod fishing . . . with occasional sealing by trap-nets. August: advent of dog-fish, end of cod fishing, washing and drying of cod, repairing of herring nets, berry picking. September: herring fishing, salting of cod, beginning of wood gathering. October: wood gathering with dog teams. December to March: fox and weasel trapping. March: sealing rice flour. April to May: sealing by trap-nets, sealskin curing and sealhide "barking."[19]

Planning in various cultures differs not only in relation to environmental and economic prerequisites but also in relation to other cultural criteria. In the "Magical Universistic" book of chronomancy, the Taoists prescribe "the propitious days on which to contract marriages, or remove to another house, or cut clothes; days on which one may begin works of repair of houses, temples, ships."[20] Islam designates Monday, Wednesday, Thursday, and Friday to be fortunate days and Tuesday, Saturday, and Sunday to be evil days. The Greeks had their own list of appropriate and inappropriate days for planning various activities. The fourth and twenty-fourth days of the month were viewed as dangerous times to engage in certain enterprises, the sixteenth day was considered unlucky for scheduling marriages, and the fourteenth day was seen as the ideal day to break in cattle.[21]

Our contemporary Western culture is so utterly obsessed with timetables and deadlines, due dates and expiration dates, that we would find it difficult, if not impossible, to tolerate the planning orientations of other cultures. Consider the simple act of planning the building of a house. Our planning is dependent on market fluctuations, labor costs, mortgage rates, inventories, and zoning regulations. Given these restraints, the goal is to build the house in the shortest time possible, at the least cost, and with the least expenditure of energy.

The Pueblo Indians plan their construction by means of an entirely different set of priorities. Before the ground is even broken, "All the right thoughts must be present."[22] The Pueblos believe that thoughts are alive and become "an integral part of any man-made structure."[23] Thoughts are as important to the construction of the building as the other materials; therefore, bad thoughts could mean the construction of a bad building. Assuring that the right thoughts will come together is the responsibility of the entire community. Each building represents the group and must not be scheduled for construction until a consensus emerges as to whether the right thoughts have come together. Such consensus could take months or years. The Pueblos are more than willing to wait, regardless of how long the process takes. Planning of construction for them is intimately bound up with group feelings and sensitivities, as opposed to marketplace planning that has more to do with expediency, efficiency, profit, and utility.[24]

Of course, we would not be able to plan activities were it not for the fact that most things repeat themselves with some measure of regularity. Rhythmic recurrence is an essential temporal dimension. We set aside certain times for specific activities and then engage in those same activities again after an established interval of time has elapsed. Every Sunday we go to church. On April 15 we pay taxes. Every six months we take our child to the dentist for a checkup.

Societies differ markedly in how they establish rhythmic regularity. In agricultural societies, the rhythms of communal life follow closely the rhythms of nature. The rising and setting sun, the cycles and seasons of nature, the periodicities of the biological and physical environment all condition the repetitive rhythms of the social order. The modern urban environment creates an entirely different set of artificially recurring rhythms. Morning and evening rush hours, assembly line production, shift work schedules, and the like establish a repetitive and highly predictable rhythmic pattern, divorced from the rhythms of the natural world.

Over a century ago, Henry David Thoreau took note of the dramatic change in rhythmic regularity that was occurring as America made the transition from an agrarian to an industrial way of life. Thoreau mused over the profound temporal impact the railroad was making on the social life of the village community:

> I watch the passage of the morning cars with the same feeling that I do the rising of the sun, which is hardly more regular. . . . The startings and arrivals of the cars are now the epochs in the village day. They go and come with such regularity and precision, and their whistle can be heard so far, that the farmers set their clocks by them, and thus one well-conducted institution regulates a whole country.[25]

Anyone who has ever traveled extensively is also acutely aware of the differences in pace that exist between various cultures. Every society has its own unique tempo, and nothing so captures the American tempo as the word "speed." We are a nation in love with speed. We drive fast, eat fast, make love fast. We are obsessed with breaking records and shortening time spans. We digest our life, condense our experiences, and compress our thoughts.

We are a culture surrounded by memos and commercials. While other cultures might believe that haste makes waste, we are convinced that speed reflects alertness, power, and success. Americans are always in a hurry.

In our educational system, a premium is placed on how fast we can recite an answer or solve a problem. Pondering, reflecting, and musing might well be encouraged in other cultures but play little or no role as modes of thought in the American educational system. Keeping up requires quick absorption of material and even faster recall. Children are taught to compete with the clock in classrooms across the country. Exams are cued to time deadlines and achievement is measured by how many answers can be completed in the time allotted. Our society is unwavering in its belief that intelligence and speed go together and that the bright child is always the fastest learner.

One of the teacher's primary responsibilities is to establish a pace and rhythm in the classroom that mimics the tempo in the larger world for which the children are being prepared. Teachers entrain students to "keep up" with their work, keep up with each other, and keep ahead of the game. Students are taught to cram, compartmentalize, and segment their learning to conform with the dictates of clocks, bells, and schedules. Even the pace and tempo in the hallways, as students move to and from classes, comes to resemble the frenetic and often frantic rhythms of the larger urban environment.

Psychologists Robert Knapp and John Garbutt conducted a study with seventy-three male undergraduates and found that the highest achievers on standardized tests were those who placed a high value on the notion of speed. The students were asked to read through a series of phrases and list by order of preference the five expressions that evoked the most satisfying images of time, followed by the five that expressed the next most satisfying images of time, and, finally, the five that expressed the least satisfying images of time. The images of time ranged from fast-moving to static. When the students' lists were correlated with their achievement, researchers found that those who chose fast-moving images of time also performed best on standardized tests while those who chose slow-moving or static images of time performed worst.

Knapp and Garbutt identified three major clusters of students in the study. The first cluster was called the Dynamic-Hasty group. They chose the fastest-moving images of time and were the highest achievers on the standardized tests. They preferred to think of time as a dashing waterfall, a speeding train, a fast-moving shuttle, a galloping horseman, a fleeing thief, a spaceship in flight, or a whirligig. The second cluster of students was identified as Naturalistic-Passive. Their images of time were drawn from nature and suggested little or no movement. They viewed time as a vast expanse of sky, a quiet, motionless ocean, a road leading over a hill, drifting clouds, wind-driven sands, the Rock of Gilbraltar, and budding leaves. The third cluster, identified as Humanistic, viewed time in terms of human surrogate figures and artifacts. Time was seen as a string of beads, a winding spool, a burning candle, an old woman spinning, an old man with a staff, a devouring monster, a tedious song, and a large revolving wheel.[26]

The Dynamic-Hasty group expresses a view of time we have come to pay homage to during the industrial age. Time is seen as linear, fast-moving, ever accelerating, and a scarce resource. The Naturalistic-Passive cluster is more reminiscent of Eastern thought and the time perspective in traditional agrarian and pastoral cultures. Time is seen as cyclical, repetitive, and sacred. The Humanistic cluster is closely related to the kind of time sense that dominated classical Mediterranean thought.

The high achievers see time as an obstacle to overcome, and an enemy to defeat. They equate faster and faster learning with victory over time; to win is to beat the clock. Our testing system, indeed our entire educational system, penalizes those students who view time in more passive, natural forms. These kinds of students are likely to be more open and vulnerable, and less controlling and manipulative in their approach to learning. They are apt to view life more as an aesthetic experience than a contest. They learn more by participation than by detachment. They are less quantifiable and, while they tend to be more artistic, are often less articulate.

Anthropologist Irving Hallowell points out that the very idea of measuring speed is derivative of the modern frame of mind and would have been unthinkable in a preindustrial world where time was not yet graduated into small, easily quantifiable temporal units. Hallowell notes that when psychological tests, "standardized with respect to speed in performance," are given to natives of nonindustrial cultures, they perform poorly. In large measure this is because speed does not hold as much value for these people when they answer questions.[27]

Speed is a temporal notion born, weaned, and nurtured in the industrial setting. In only a few short centuries, "it has risen into prominence as a value of Western society and functions as an important factor in the motivation of individuals."[28] Historian William Durant once remarked that "no man in a hurry is quite civilized."[29] If we were to take Durant's statement at face value, we might have to rethink seriously the long-standing conviction that the industrial way of life and human progress go hand in hand.

Even among urbanized people, where speed is a universally held value, marked differences in tempos exist between cultures. Several years ago researchers compared six countries in terms of pace of life and found that each nation did indeed possess its own unique temporal rhythm. The researchers examined three time indicators: the accuracy of the country's clocks, the speed of pedestrian traffic, and the time it took for a postal clerk to sell a stamp. The study was conducted in the largest city of each country and in a medium-size urban area. The countries studied were Japan (Tokyo and Sendai); Taiwan (Taipei and Tainan); Indonesia (Jakarta and Solo); Italy (Rome and Florence); England (London and Bristol); and the United States (New York City and Rochester).

It should come as no surprise that Japan's clocks were by far the most accurate, averaging less than a half-minute early or late. Indonesia, on the other hand, fared least well, with its clocks over three minutes early or late at any given moment. Japan also led the way in pedestrian speed. The average Japanese citizen walked one hundred feet on a downtown street in less than 20.7 seconds. The English came in second, averaging 21.6 seconds, with the Americans a close third at 22.5 seconds. Again, Indonesia was last, clocking in at 27.2 seconds. When it came to speed of postal service, Japan again led the pack, with clerks averaging 25 seconds to complete a transaction. The Italians came in last, taking 47 seconds on the average to complete the task.

Overall, the study showed a strong correlation between a country's clock accuracy, pedestrian speed, and postal service. Each country's pace in one category more or less correlated with its pace in the other two categories, suggesting a consistent and unique pace of life within each culture.[30]

There is yet another temporal factor indispensable to social interaction. No group, community, or society could maintain itself with any measure of cohesion were it unable to synchronize the activities of its members. Synchronizing behavior is critical to every social function, be it recreational or religious, economic or political.

Anthropologist Edward T. Hall recounts an experiment conducted by one of his students that illustrates the tremendous hidden power of synchronization in the coordination of social activity.

The student hid in an abandoned car next to a schoolyard and, from that location, filmed children playing with each other at recess. After playing back the film at several speeds, he noticed one "very active little girl who seemed to stand out from the rest."[31] This particular child would flit between various clusters of children throughout the playground. Upon closer observation, the film showed that whenever the little girl came near a group of children they would begin to synchronize their behavior with each other and with her. After many viewings the researcher concluded "that this girl, with her skipping and dancing and twirling, was actually orchestrating the movements of the entire playground."[32]

The pattern of the little girl's movement was very much like a silent dancebeat. There was a rhythm to it that the researcher had come across before. It turned out that the rhythmic pattern that the little girl had established in the playground was identical to a particular piece of rock music that was popular at the time. When the music was overlaid onto the film, "the entire three and a half minutes of the film clip stayed in sync with the taped music! Not a beat of the frame was out of sync."[33]

Social interaction at every level requires synchronization. In cultures like our own, where a high premium is placed on personal autonomy, synchronized activity is more an expression of many individual human beings working in tandem to accomplish a common goal. We tend to see ourselves in Newtonian terms, as discrete material entities interacting with other discrete material entities. For synchronization to be effective within this kind of atomistic framework, it is necessary to establish, in advance, well-thought-out rules and regulations in order to coordinate properly the interactions of individuals, each of whom has his or her own priorities and needs.

In other cultures where the collective will of the community takes precedence over the individual will of its members, synchronizing activity comes somewhat easier as people perceive themselves more in organismic rather than individualistic terms. Because these cultures already think in terms of wholes, they do not have to take extraordinary measures to get in step with one another. The group is not so much an instrumental unit born of negotiation and compromise between its individual members, but more of a biological unit that exists a priori, independent of the tasks it takes on.

The differences between the organismic and individualistic approaches to synchronized activity is best illustrated in the case of the two most highly industrialized countries in the world, Japan and the United States. In the United States the individual takes precedence. In Japan the group does. Nowhere is this more in evidence than in the workplace. American corporate executives have been surprised and impressed by the group dynamic in Japanese factories. For example, at the Toyota automobile factory, the "assembly line teams start the day doing exercises together, then they work together, take their breaks together, eat together, live next to each other in a company compound, and even vacation together."[34]

The Japanese are much more aware of group dynamics and pay much greater attention to synchronization than do Americans. Even when it comes to social conversation, the Japanese will often "monitor their breathing in order to stay in sync with their interlocutor."[35] American businessmen are just now coming to understand how important this kind of organismic synchronization is to securing a competitive advantage in the international marketplace.

Of all the temporal dimensions, none has pricked the curiosity of sociologists more than time perspective. American culture has always been more present- and future-oriented in its temporal perspective. Our frontier origins have predisposed us to living from moment to moment. In the precarious process of taming a continent, we have had to learn how to concentrate our energies on day-to-day survival. Because we are a nation of pioneers, we are imbued with the notion that we must keep on moving and never look back.

We do not spend a great deal of time on ritualizing the past. We prefer novelty to tradition and are in love with anything that is young, new, or unexplored. If we are not very reflective, it is also the case that we do not extend our temporal horizon very far into the future. We are primarily interested in the immediate future, the future that can be expropriated tomorrow or the next day, but are little concerned with time spans that reach beyond our limited lifetimes.

Other cultures are much more past- or future-oriented. Some cultures have successfully stretched their temporal perspective in both directions, encompassing the ancient past and the unforeseen future. The Iroquois Nation is exceptional among the cultures of the world in this regard.

Every decision that comes before the Iroquois chieftains is subjected to a rigorous temporal examination. The Iroquois see themselves as servants of the past and stewards of the future. Their ancestors counsel them from the grave just as their unborn children cry out to them from somewhere beyond the horizon.

Decision making, for the Iroquois, is an affair that reaches far beyond the moment and the limited concerns of those huddled around the campfire. When the Iroquois make decisions, they do so always with the thought of honoring their ancestors and nurturing their unborn progeny. They ask: How does the decision we make today conform to the teachings of our grandparents and to the yearnings of our grandchildren?

The Iroquois are unique among cultures in that they have institutionalized a specific future time frame into all decision making. An Iroquois chief explains the process:

> We are looking ahead, as is one of the first mandates given to us as chiefs, to make sure [that] every decision we make relates to the welfare and well-being of the seventh generation to come, and that is the basis by which we make decisions in council. We consider: Will this be to the benefit of the seventh generation? This is a guideline.[36]

American policy leaders might find such an exercise difficult to even imagine. Their concept of decision-making responsibility barely extends beyond the four-year period that marks off each new general election.

This difference in time perspective between the Iroquois Nation and contemporary American society is illustrative of the tremendous variation that exists in temporal orientation between cultures. Every society imprints its own unique temporal design. The six primary temporal dimensions of sequence, duration, planning, rhythm, synchronization, and time perspective have been constructed and assembled in myriad ways by various cultures, providing a rich diversity of anthropological time zones for the human family to dwell in.

Although temporal orientations have varied substantially between cultures, every society entrained its social clocks, at least partially, to the biological and physical rhythms of the natural world. With the dawn of the industrial age, however, civilization began accelerating the process of separating itself from the time orientation of the planet. Today, that separation is nearly complete. To understand the historical process that has led to today's simulated time world, it is necessary to examine the impact that various time-allocating devices have had on the temporal dynamics of Western culture.

PART II

DIVIDING THE TIME PIE

4

CALENDARS AND CLOUT

I n examining the similarities and differences in temporal orientations between cultures, it becomes strikingly apparent
how important time is in defining the character of a society and
a people. Emile Durkheim, one of the founders of modern sociology, sounded a profound note when he proclaimed time to be
the centerpiece of social life. "Societies organize their lives in
time and establish rhythms that then come to be uniformly imposed as a framework for all temporal activities."[1] Another of the
great social scientists of the early part of the twentieth century,
Pitirim Sorokin, agreed with Durkheim that time is an essential
category of social experience. In his classic essay, "Sociocultural
Time," Sorokin made clear that he considers time measurement
to be the critical factor in social relations:

> The possession of means and ways to "time" the behavior of
> the members of any group in such a way that each member
> apprehends "the appropriate time" in the same way as do other
> members has been possibly the most urgent need of social life
> at any time and at any place. Without this, social life itself is
> impossible.[2]

The human race has relied on four major time-allocating devices throughout history: biotic rituals, astronomical calendars,
clocks and schedules, and now computer programs. With the

introduction of each new device, the human species has detached itself further from the biological and physical rhythms of the planet. We have journeyed from close participation with the tempo of nature to near isolation from the earth's rhythms.

Through most of prerecorded history, our ancient ancestors calculated time exclusively by reference to natural phenomena. The migratory rhythms of the great herds of animals and the gestation and ripening times of wild herbs and roots provided the necessary temporal markers for ordering the social life of Paleolithic tribes. Hunter-gatherers institutionalized their close temporal bond with ecological events by the enactment of sacred rituals. Their repertoire of dances and chants was designed to accompany the important succession of seasonal changes around them.

As human societies metamorphosed from hunter-gatherer to agricultural economies, interest shifted from biological time clocks to cosmic time clocks. By observing the changing constellations of planets and stars in the heavens, human beings were able to develop a far more sophisticated time-reckoning system. The shifting emphasis from biotic to cosmic clocks was accompanied by a new shift in time-ordering devices. Sacred rituals, which had long been used to order the temporal affairs of hunter-gatherer societies, were joined by the birth of the calendar, a new time-ordering device more suited to a sedentary agricultural form of existence.

It is through the use of calendars that advanced civilizations have created reference points for shared group activity. The concept of days, weeks, months, years, the celebration of births and the memorializing of deaths, the recording of changing seasons, and the acknowledgment of rites of passage, are all incorporated in and an outgrowth of the creation of calendars.

Until the modern era, most calendars were woven from a blend of religious, environmental, and economic considerations. Today, the holy days and seasonal markings that once dominated the calendars and the social life of every culture must now compete with other temporal reference points such as quarterly income statements, fiscal years, five-year plans, ten-year national census reports, and political celebrations including May Day, Memorial Day, and Labor Day.

Whether sacred or secular, every calendar expresses the essential politics of a culture. No other device in the entire political repertoire is as critical as the calendar in forging a sense of group cohesion.

The Jewish experience provides a good example of the importance of calendars in maintaining group identity and preservation. The Jews have relied on their own calendar for close to four thousand years. Jews throughout the world today continue to use their ancient calendrical system alongside the secular calendars of their contemporary culture. Israel uses the Jewish calendar to regulate the activities of the nation.

For most of their existence, Jews have been a nomadic people. Their common bond has been less geographical than spiritual, less territorial than temporal. In the early days, they were forced to live as slaves in the land of the pharaohs. After the destruction of the temple in Jerusalem and the sack of Rome, the Jews were scattered across Europe in small bands, isolated from each other. Even as they remained physically separated, they looked to their calendar as a way of sharing group experiences. It has been said of the Jewish calendar that it has "united all those who have been scattered around the world and made them one people."[3]

In recent years Jewish scholars have agonized over the increasing movement away from the Jewish calendar and the assimilation into the secularized Christian calendar that now dominates much of the social life of the planet. Hebrew scholar Joshua Monoach argues that:

> The soul of Israel, its religion and its customs, is anchored in
> its time. Replacing its national-religious time by the time of
> others . . . is suicidal for a distinct and independent people.[4]

Monoach warns his fellow Jews that the abandonment of their national calendar could spell the end of their culture. He notes that "every people that has tried to separate itself from its time has disappeared and is no longer remembered among the living."[5]

Of all the Jewish calendrical innovations, none is more poignant than the creation of the Sabbath. Among all of the Jewish time markings, it remains the most hallowed and the most honored observance. The Sabbath is a testimonial to the power that calendars exercise over the life of a people. Through the Sabbath ritual, Jews have effectively been able to separate themselves from the rest of the world around them and preserve, intact, both their heritage and their shared future vision.

In the Bible it is written that God created the world in six days and rested on the seventh. It is the seventh day that the Sabbath celebrates as once every week Jews around the world remove themselves from secular-world time and enter into sacred time. It is a time they share together with their God, even as they remain separated from each other over great expanses of geography.

The Sabbath is a holy day and meant to be experienced as a day of pleasure. As God rested on the seventh day, so too must his "chosen people." As God reveled in his creation, so too must his flock. On this special day of rest and enjoyment, Jews redirect their entire temporal orientation. The dress code, the dietary rules, and the manner of bathing are abruptly changed on this holiest of days. Even the manner of walk is reoriented to reflect a different temporal rhythm and pace. A Jew is expected to leave behind the hectic pace of daily life and walk in a more leisurely, carefree manner.

On the Sabbath a Jew severs his relationship to secular time and the daily routine that accompanies it and enters into a timeless other world to reconsecrate the pact, or covenant, with the creator. This calendrical convention serves as a revolutionary act. Through its observance, the Jew is able symbolically to overthrow the temporal order of whatever culture he or she is in and take part in another realm of time altogether. On this day the secular culture is powerless to impose its temporal will on the life of the Jew.

The Romans discovered just how tenacious the Jewish commitment to the Sabbath could be when they were forced to exempt Jewish men from military service because they refused to work or fight on the Sabbath.

Through the Sabbath observance, Jews everywhere rekindle their own communal identity even in the midst of alien cultures wedded to very different values. The Sabbath stands as the oldest and most effective form of institutionalized rebellion in Western experience.[6]

While the calendar has always figured prominently in Jewish community life, it has also played an equally significant role in the rise of Christianity. The Church's influence over the economic and political life of Europe was considerable during the medieval centuries, and its calendrical innovations, in no small measure, helped to formalize its control over the powers and principalities.

Today the most obvious manifestation of Christian calendrical power remains the dating of history into two great eras, B.C. and A.D. Yet this was a rather late development in the history of the Church. Although first proposed as a method of dating in A.D. 525 by a monk and mathematician named Dionysius Exiguus, this reform did not take hold until centuries later.[7]

From the beginning, Christian calendrical reforms were steeped in the politics of the day. Consider the dating of Christmas. Originally the Nativity was celebrated on January 6, along with Epiphany. The Church was anxious to separate the two events, realizing that the birth of Christ and the Epiphany could hardly have taken place on the same day. Consequently, the pope chose December 25 as the official commemoration of the Virgin Birth.

As it turns out, however, his choice was politically inspired. Both December 25 and January 6 marked the celebration of winter solstice rites. The Church was determined to confront these pagan rituals head-on by superimposing the Nativity and the Epiphany on the same days.[8]

Perhaps the most important calendrical controversy faced by the early Church revolved around the proper dating of Easter.

Church fathers were embarrassed by the fact that the most important event celebrated in the Christian world took place on one of the most important Jewish holidays, Passover. During the early days of Christianity, the two events were often joined together. Christians referred to the period leading up to the Easter celebration as the Week of the Unleavened Bread.

The prelates of the Church worried that the close affinity of the two religious events would undermine their efforts to establish Christianity as a separate and unique religious force. Determined to free Christianity from the "Jewish influence," Church officials from all over the European continent assembled at the first ecumenical council at Nicea in A.D. 325 to resolve the problem. A resolution was passed declaring that Easter should be observed "on the Sunday following the full moon which coincides with, or falls next after, the vernal equinox." Since Passover always coincides with the full moon, this calendrical reform assured that the two holy days would never again coincide.[9]

This calendrical change marks a major turning point in Church history. While the old guard was intent on maintaining apostolic tradition, the reformers were anxious to "emancipate" the Church from its Jewish parentage once and for all. The modernists were led by Emperor Constantine who, more than any other individual, was responsible for transforming Christianity from a small sect into the institutionalized religion of Europe. Constantine was well aware of the political significance of dissociating Christianity from Judaism and said as much in a letter he issued to the faithful after the calendrical reform was passed. In the text of the communiqué, Constantine remarked:

> It appeared an unworthy thing that in the celebration of this most holy feast we should follow the practice of the Jews. . . . For we have it in our power, if we abandon their custom, to prolong the due observance of this ordinance to future ages. . . . Let us then have nothing in common with the detestable Jewish crowd.[10]

The Passover-Easter controversy is a graphic historical illustration of the intimate relationship that exists between calendars and group identity. Noting that what binds people together as a community is their shared temporal order, Eviatar Zerubavel concludes:

> The calendar helps to solidify in-group sentiments and thus constitutes a powerful basis for mechanical solidarity within the group. At the same time, it also contributes to the establishment of intergroup boundaries that distinguish, as well as separate, group members from "outsiders."[11]

The most radical attempt at redirecting the temporal identity of an entire culture through calendrical manipulation took place during the French Revolution. The architects of the Revolution were committed to ridding Western civilization of what they considered to be the "religious superstition," cruelty, ignorance, and oppression of the church and state reign of the previous era. They entertained a new image of the future in which reason would reign as the cardinal virtue and provide the context for their utopian vision. To advance this revolutionary goal, the leaders of the new French Republic issued sweeping reforms in the social, economic, and political life of the society and then attempted to institutionalize these changes by transforming the entire temporal frame of reference of the French people.

On November 24, 1793, the National Convention of revolutionary France put into effect a radical new calendrical system, one that reflected the ideals and principles of the new revolutionary regime. Their motivation was clearly political. They realized that as long as the Christian calendar was allowed to remain the primary temporal reference, it would be impossible to rid French culture of the prerevolutionary influence of the Church. In order to consolidate their gains and ensure that the French people would not fall prey to counterrevolutionary tendencies, it was deemed necessary to obliterate all temporal references that might, in any way, perpetuate loyalty to the past. Writing on the political significance of adopting an entirely new calendar, political scientist Thomas Darby observes:

> Its institution was an extreme, yet subtle method designated to eradicate the popular consciousness of all previous associations, loyalties, and habits, and to substitute in their stead new ones that emphasized the revolutionary ideology. This was to have the doubled effect of first instituting a state of "mass forgetfulness" and second of inaugurating the founding of a new popular memory.[12]

The new French calendar was designed in part to de-Christianize time, that is, to eliminate the Church's influence over the time of the French people. It was also intended to inculcate a new temporal awareness in which the values of secularism, rationalism, naturalism, and nationalism would determine the sequencing, duration, scheduling, coordination, and temporal perspective of the new French man and woman. Eviatar Zerubavel sums up the revolutionary intent of the new calendrical decree:

> The abolition of the traditional temporal reference framework was deliberately meant to strip the Church once and for all of one of its major mechanisms of exercising social control and regulating social life in France.[13]

The new calendar replaced the Christian era with the Republican era. Instead of using the birthdate of Christ as the dividing point between ancient and modern history, they chose to substitute the birthdate of the French Republic. Henceforth, 1792 was to be looked on as Year One in what the calendrical architects viewed as a new age in history.

In their zeal to be as rational and scientific as possible, the new government redesigned the entire year to conform with the decimal system. The revolutionary calendar was composed of twelve months, each containing thirty days. Each month, in turn, was divided into three ten-day cycles called decades. Each day was divided into ten hours, and each hour further divided into one hundred decimal minutes. Each minute contained one hundred decimal seconds.[14]

The days of the week were renamed using numbers only: Day One, Day Two, Day Three, and so forth. The designers were mindful of the long history of religious significance attached to the traditional names used to demarcate the week. In substituting pure numbers with mathematical significance only, they believed they were advancing the interest of reason over religious superstition.[15]

The new calendar also attempted to reestablish a relationship with the natural environment. All of the saints' days and holy days in the old Christian calendar were abolished, and in their place were substituted natural phenomena. Instead of honoring a saint, the new French man and woman was expected to honor a particular tree, plant, animal, or flower. Even the names of the months were changed to reflect the interest in aligning the new era with the rhythms of nature. The new months were named Vintage, Mist, Frost, Snow, Rain, Wind, Seeds, Flowers, Meadows, Harvest, Heat, and Fruits.[16]

The French public did not take kindly to the elimination of the holy days. Under the old Christian calendar, the Church laid aside fifty-two Sundays, ninety rest days, and thirty-eight holidays. The new calendar eliminated Sundays and all of the rest of the holidays, leaving the French citizen with nothing to look forward to but a regimen of never-ending work. To compensate, the calendar designers introduced a limited number of special rest days with names like The Human Race Day, The French People Day, The Benefactors of Mankind Day, Freedom and Equality Day, The Republic Day, Patriotism Day, Justice Day, Friendship Day, Conjugal Fidelity Day, and Filial Affection Day. Needless to say, by reducing the number of rest days from more than 180 to 36, the French Republic insured the animosity of the French people.[17]

The French revolutionary calendar lasted only thirteen years.

In 1806 Napoleon reinstated the Gregorian calendar, in part to appease the French public, who had resisted the revolutionary calendar from the beginning, and in part to placate the pope in the hopes of entertaining a rapprochement with the Vatican. The new calendar had been doomed from the outset. In attempting to obliterate every temporal benchmark in the life of the French people, the architects of the new calendar created the ideal condition for reaction, retrogression, and, inevitably, repudiation.

5

SCHEDULES AND CLOCKS

For much of recorded history the calendar ruled over human affairs. It served as the primary instrument of social control, regulating the duration, sequence, rhythm, and tempo of life and coordinating and synchronizing the shared group activities of the culture. The calendar is past-oriented. Its legitimacy rests on commemoration. Calendar cultures commemorate archetypical myths, ancient legends, historical events, the heroic deeds of gods, the lives of great historical figures, and the cyclical fluctuations of astronomical and environmental phenomena. In calendar cultures, the future takes its meaning from the past. Humanity organizes the future by continually resurrecting and honoring its past experiences.

The calendar continues to play an important role in contemporary culture. Its political significance has been greatly reduced, however, with the introduction of the schedule. The schedule exerts far greater control over time allocation than the calendar. While the calendar regulates macro time—events spread out over the year, the schedule regulates micro time—events spread out over the seconds, minutes, and hours of the day. The schedule looks to the future, not the past, for its legitimacy. In scheduling cultures, the future is severed from the past and made a separate and independent temporal domain. Scheduling cultures do not commemorate; they plan. They are not interested in resurrecting the past but in manipulating the future. In the new time frame, the past is merely prologue to the future. What counts is not what was done yesterday, but what can be accomplished tomorrow.

The calendar and the schedule differ in still another important way. While modern calendars have become increasingly secularized, throughout most of history their social content was inseparably linked to their spiritual content. In traditional calendrical cultures the important times are sacred times and are observed through the commemoration of special holy days. The schedule, in contrast, is associated with productivity. Sacred values and spiritual concerns play little or no role in the formulation of schedules. Time, in the new scheme of things, is an instrument to secure output. Time is stripped of any remaining sacred content and transformed into pure utility.

George Woodcock has observed, "It is a frequent circumstance of history that a culture or civilization develops the device that will later be used for its destruction."[1] The schedule, more than any other single force, is responsible for undermining the idea of spiritual or sacred time and introducing the notion of secular time. Ironically, the revolution in time wrought by the introduction of the schedule begins with the Benedictine monks.

The Benedictine order was founded in the sixth century. It differed, in one important respect, from other church orders. St. Benedict emphasized activity at all times. His cardinal rule, "Idleness is the enemy of the soul," became the watchword of the order.[2] The Benedictines engaged in continual activity, both as a form of penitence and a means of securing their eternal salvation. St. Benedict warned the members of his order that "if we could escape the pains of hell and reach eternal life, then must we—whilst there is still time—hasten to do now what may profit us for eternity."[3]

Like the merchant class that would follow in their shadow, the Benedictines viewed time as a scarce resource. But for them time was of the essence because it belonged to God, and, because it was his, they believed they had a sacred duty to utilize it to the fullest in order to serve his glory. Toward this end, the Benedictines organized every moment of the day into formal activity. There was an appointed time to pray, to eat, to bathe, to labor, to read, to reflect, and to sleep. To ensure regularity and group cohesiveness, the Benedictines reintroduced the Roman idea of the hour, a temporal concept little used in the rest of medieval society. Every activity was assigned to an appropriate hour during the day. Consider the following set of instructions from the Rule of St. Benedict:

> The brethren . . . must be occupied at stated hours in manual labor, and again at other hours in sacred reading. To this end we think that the times for each day may be determined in the following manner. . . . The brethren shall start work in the morning and from the first hour until almost the fourth do the tasks that have to be done. From the fourth hour until the sixth let them apply themselves to reading. After the sixth hour, having left the table, let them rest on their beds in perfect silence.[4]

Under the Rule of St. Benedict, even bodily functions were made to conform with this new, highly structured temporal order:

> The brethren shall rise at the eighth hour of the night so that their sleep may extend for a moderate space beyond midnight. . . . Let the hour of rising be so arranged that there be a short interval after matins, in which the brethren may go out for the necessities of Nature.[5]

To make sure that everyone began each activity together at the prescribed moment, the Benedictines introduced bells. Bells pealed, jangled, and tinkled throughout the day, hurrying the monks along to their appointed rounds. The most important bells were those that announced the eight canonical hours when the monks celebrated the Divine Offices.

The Benedictines ordered the weeks, and the seasons, with the same temporal regularity as they did the day. Even such mundane activities as head shaving, bloodletting, and mattress refilling took place at fixed times during the course of the year.[6]

The very idea of an appointed time for each activity was revolutionary. The Benedictines elevated this radical new concept to the level of a moral principle. Strict adherence to this exacting temporal orientation was considered commendable in the eyes of the Church and, they believed, in the eyes of the Lord as well. In fact, so seriously was this new approach to time taken that the monks were instructed to sleep with their clothes on, as "being clothed they will thus always be ready, and rising at the signal without any delay may hasten to forestall one another to the work of God."[7]

Of course, this kind of surrender to the dictates of the hour and its proscribed activity ensured that each monk's time would be given over to the institution and its guardians. The individual monks were locked into a temporal order so rigidly defined that there was no time left for individual initiative. In this way, the monastery predated the autocratic state by nearly a millennium.

The Benedictines introduced more than a new temporal orientation when they introduced the "schedule." Eviatar Zerubavel wisely observes that, in appointing prescribed hours for specific activities and in demanding rigid obedience to the performance of these activities at the appropriate time, the Benedictines "helped to give the human enterprise the regular collective beat and rhythms of the machine."[8] Political scientist Reinhard Bendix has described the Benedictine monk as "the first professional of Western Civilization."[9]

To secure proper compliance with the prescribed schedule, the Benedictines developed a tool that could provide them with greater accuracy and precision of time measurement than could be obtained by reliance on bells and bell ringers. They invented the mechanical clock. Lewis Mumford once remarked that "the clock, not the steam engine, is the key machine of the Modern Age."[10]

The first automated machine in history ran by a device called an escapement, a mechanism that "regularly interrupted the force of a falling weight," controlling the release of energy and the movement of the gears.[11]

At first this new invention was used exclusively by the Benedictines as a means of assuring greater conformity with the daily schedule of duties. The clock allowed the clergy to standardize the length of hours. By establishing a uniform unit of duration, the monks were able to schedule the sequence of activities with greater accuracy and synchronize group efforts with greater reliability.

It was not long, however, before word of the new marvel began to spread. By the late fifteenth century, the mechanical clock had stolen its way out of the cloisters and had become a regular feature of the new urban landscape. Giant clocks became the centerpiece of city life. Erected in the middle of the town square, they soon replaced the church bell as the rallying point and reference point for coordinating the complex interactions of urban existence.

Just a century earlier, the grandeur of the Gothic cathedral had marked the status of a community, but now the erection of the town clock became the symbol of city pride. In 1481, the residents of Lyons petitioned the city magistrate for a town clock, justifying the expenditure of city funds on the grounds that "more people would come to the fairs, the citizens would be very contented, cheerful and happy, and would live a more orderly life."[12]

The construction of these giant public clocks was no small affair, as the following account attests: On January 15, 1356, master clockmaker Anthony Bovell was commissioned to construct a clock for King Peter IV of Aragon. The clock was to weigh two tons and the bell that accompanied it was to weigh four tons. A heavily reinforced tower had to be constructed to house the clock and bell. To accomplish this task, Bovell had three furnaces built and hired ten more clockmakers, in addition to an army of masons, ropers, founders, bricklayers, carters, plasterers, and smiths totaling nearly a hundred. Local merchants were con-

tracted to supply materials and equipment. After months of forging and casting the iron fittings and additional months building
the tower itself, Bovell and his workmen had to rig together a
mobile derrick and huge cranes to hoist the clock and bell into
place atop the tower. The cost of the entire project was enormous
by the standards of the day, but deemed well worth it by the
Court, which could now boast of having an automatic device to
regulate the princely and commercial affairs of the kingdom.[13]

The most famous of the early clocks was one constructed at
Strasbourg. The Senate of Strasbourg contracted with Conrad
Dasypodius, a professor of mathematics at Strasbourg Academy,
to build a "magnificent, splendid, and artistic work" that would
"bring honor to the Senate and the people of Strasbourg."[14] Begun
in 1547, it took twenty-seven years to complete construction. An
extraordinary accomplishment, the clock instantly became the
talk of Europe. The largest of its kind ever built, it measured
twenty-five feet in width at the base and was over sixty feet high.
More than a simple timekeeper, the clock had built into its artifice
a compendium of automatic devices and automated puppetry
laced and adorned with beautiful works of art.

Dasypodius had surrounded his masterclock with the combined knowledge of the Western world, making the clock itself
the symbolic center of Western consciousness.

> The automated astronomical devices included a large calendar
> dial with the holy days, a clock giving local time, an astrolobe
> with delicately wrought signs of the zodiac and planets, a
> mechanism that showed the phases of the moon, and a ce-
> lestial sphere supported by a pelican mounted on the floor in
> front of the clock. The principal trains of automatons were:
> the tutelary gods of the days of the week being borne around
> in elaborate chariots; figures of the four ages of Man that
> struck the quarter hours in their circuit; and the figures of
> Christ and Death dueled at the stroke of the hour with Death
> winning all hours except the last. Also the old restored cock
> was mounted on the crest of the weight tower and at midday
> flapped its wings, raised its head, opened its beak, shook its

tail and crowed, to the accompaniment of a carillon playing
bits of music.[15]

The first clocks had no dials. They merely sounded a bell on
the hour. Indeed, the term clock comes from the Middle Dutch
word *clocke*, which means bell. By the sixteenth century, clocks
were chiming on the quarter hour and some were being constructed with dials to demarcate the passing of each hour. In the
mid 1600s the pendulum was invented, providing a much more
exacting and reliable timing mechanism. Shortly thereafter, the
minute hand was introduced. The second hand did not make its
debut until the early 1700s, when it was first used by astronomers, navigators, and doctors to record more accurate measure-
ments. While the idea of minutes and seconds had been conceived
back in the fourteenth century by mathematicians, it is important
to bear in mind that they did not become part of the temporal
consciousness of Western man and woman until they found their
way onto the dial of the mechanical clock.[16]

The idea of organizing time into standardized units of hours,
minutes, and seconds would have seemed strange, even macabre,
to a peasant serf of medieval times. A day then was roughly
divided into three sectors: sunrise, high noon, and sunset. The
only other reminders, says Lawrence Wright, were "the seeding
and harvest bell that called them to work, the sermon bell and
the curfew bell."[17] Occasionally one might hear the sound of the
"gleaning bell, the oven bell when the manor oven was fired to
bake the bread, the market bell, and bells that summoned them
to feast, fire, or funeral."[18] Even in these instances, time was not
something fixed in advance and divorced from external events.
Medieval time was still sporadic, leisurely, unpredictable, and,
above all, tied to experiences rather than abstract numbers.

"By its essential nature," observes Lewis Mumford, the clock
"dissociated time from human events."[19] It is also true, as historian David Landes, of Harvard University, suggests, that the
clock dissociated "human events from Nature."[20] Time, which

had always been measured in relation to biotic and physical phenomena, to the rising and setting sun and the changing seasons, was henceforth a function of pure mechanism. The new time substituted quantity for quality and automatism for the rhythmic pulse of the natural world.

The emerging bourgeois class of merchants embraced the mechanical clock with a vengeance. It quickly became apparent that the increasingly complex activities of urban and commercial life required a method of regulation and synchronization that only the clock could provide.

The clock found its first use in the textile industry. While textile production predated the rest of the industrial revolution by two centuries, it embodied many of the essential attributes that were to characterize the coming age. To begin with, textile manufacturing required a large centralized work force. It also required the use of complex machinery and great amounts of energy. The new urban proletariat congregated each morning in the dye shops and fulling mills "where the high consumption of energy for heating the vats and driving the hammers encouraged concentration in large units."[21] This type of complex, highly centralized, energy-consuming production technology made it necessary to establish and maintain fixed hours for the beginning and end of the workday.

Work bells, and later the work clock, became the instrument of the merchants and factory owners to control the work time of their laborers. Historian Jacques Le Goff remarks that here was the introduction of a radical new tool to assert power and control over the masses. Bluntly stated, "the communal clock was an instrument of economic, social, and political domination wielded by the merchants who ran the commune."[22]

It was not only in the factory that the clock played an important new role. The bourgeois class found use for it in virtually every aspect of their daily life. This was a new form of temporal regimentation, more exacting and demanding than any other ever conceived. The bourgeoisie introduced the clock into their homes, their schools, their clubs, and offices. No corner of the culture was spared the reach of this remarkable new socializing tool. Lewis Mumford took stock of this radical transformation in time consciousness and concludes that:

> The new bourgeoisie, in counting house and shop, reduced life to a careful, uninterrupted routine: so long for business; so long for dinner; so long for pleasure—all carefully measured out. . . . Timed payments; timed contracts; timed work; timed meals; from this time on nothing was quite free from the stamp of the calendar or the clock.[23]

To become "regular as clockwork" became the highest values of the new industrial age.[24] Without the clock, industrial life would not have been possible. The clock conditioned the human mind to perceive time as external, autonomous, continuous, exacting, quantitative, and divisible. In so doing, it prepared the way for a production mode that operated by the same set of temporal standards.

The potential power of the clock was not lost on those in high places. In 1370, Charles V issued a proclamation ordering that all the bells of Paris be regulated by the clock at the Palais Royal.[25] David Landes best captures the political impact of introducing the clock into European culture:

> The invention of the mechanical clock was one of a number of major advances that turned Europe from a weak, peripheral, highly vulnerable outpost of Mediterranean civilization into a hegemonic aggressor. Time measurement was at once a sign of new-found creativity and an agent and catalyst in the use of knowledge for wealth and power.[26]

In the new world of the bourgeoisie, all of life's activities were brought under the tight control of the schedule and the clock.

The day was segmented into preplanned activities and time was divided into standardized units.

Time was snatched away from its biological and environmental moorings and locked up inside the gears of an automated machine that now parceled it out in steady, nondescript beats.

6

TIME SCHEDULES AND
FACTORY DISCIPLINE

Armed with a new time construction, the emerging bourgeois class of merchants and factory owners was now capable of effectively regulating the new industrial form of production. Converting the masses to the new time orientation became the urgent business of the age. The baptism took place at the factory gates.

The new industrial mode of production required uncritical acceptance of the new bourgeois concept of time. Convincing European labor to adopt "bourgeois time," however, proved to be a formidable task.

Medieval craftsmen had been accustomed to a very different temporal rhythm. Generally self-employed, they set their own work time. For the most part, they were task-oriented and thought nothing of suspending work for several hours during the day to chat with a friend, do chores around the house, or visit the local tavern. Since most trades were seasonal, craftsmen would work for concentrated periods of time, then engage in rest and recreation for long periods. Peasant farmers followed a similar routine, their rhythm of work and leisure dependent on the dictates of the seasons.

The new system of factory production required a wholly new temporal orientation. Large numbers of workers had to be organized under one roof. Work tasks were divided into specialized categories requiring greater coordination of activity. The new steam-powered machinery required constant attention. If one or more workers left their designated work area unattended, the whole production site might grind to a halt. Then, too, machinery was expensive to construct, install, and run. Consequently, it could not easily be turned on and off to suit the whims of workers who might prefer to show up to work late, take a break for an unspecified period of time, or leave early.[1]

Whereas in the craft trades and in farming the workers had set the pace of activity, in the new factory system the machinery dictated the tempo. That tempo was incessant, unrelenting, and exacting. The industrial production mode was, above all else, methodical. Its rhythm mirrored the rhythm of the clock. The new worker was expected to surrender his time completely to the new factory rhythm. He was to show up on time, work at the pace the machine set, and then leave at the appointed time. Subjective time considerations had no place inside the factory. There, objective time—machine time—ruled supreme.

The new time orientation was one that the masses were unaccustomed to and, for the most part, unwilling to accept. When the first cotton mills were opened in Scotland, managers complained that "the people were found very ill-disposed to submit to the long confinement and regular industry that is required of them."[2] A hosier reported that in his factory he "found the utmost distaste on the part of the men to any regular hours or regular habits. . . . The men themselves were considerably dissatisfied, because they could not go in and out as they pleased, and have what holidays they pleased, and go on just as they had been used to do."[3]

The workers so loathed the new time orientation that many factory owners were simply unable to secure a labor force. When they did, absenteeism was high and often workers would quit after just a few short weeks. In many firms it was not unusual to experience a one hundred percent labor turnover in a year.

Some groups refused *en masse* to surrender to the new factory discipline. The peasants of the Scottish highlands could not easily be made to accept the new time frame. One observer remarked that the Highlander "never sits at ease at a loom; it is like putting a deer in the plough."[4]

According to Sidney Pollard, whose book *The Genesis of Modern Management* catalogues many of the early struggles between workers and owners, most laborers could only be made to accept the new factory rhythm if destitute and desperate. For this reason, Glasgow factory owners were able to take advantage of the agricultural crisis in Ireland to recruit their work force. "They preferred the Irish," writes Pollard, because they "were docile and willing to take starvation wages on first arrival."[5]

In the late eighteenth century, a new word entered the popular vocabulary. Punctuality came from the Latin *punctus*, which originally meant "details of conduct." It now came to be associated with the moral imperative to "be on time."[6] Getting workers to work at the appointed clock hours was a recurring problem. In Lancaster as in other industrial cities, a steam whistle would blow at five in the morning to wrest people from their sleep.[7] If that proved insufficient, employers would hire "knockers up," men who went from flat to flat "rapping on bedroom windows with long poles." Some of the knockers up even pulled on strings "dangling from a window and attached to the worker's toe."[8]

Getting to work was one problem. Adhering to factory rhythms was still another. The factory was the first place that the common man and woman were exposed to the schedule. It had been several hundred years since the Benedictines had first introduced the schedule, but for most people it had remained a curiosity, a temporal aberration confined to the life of the monastery. But with industrial production and the factory system, the schedule suddenly became a very real and intimate part of the time frame of secular life. The bourgeois class relied on the schedule to plan their commercial activities. There were delivery schedules, inventory schedules, bank and accounting schedules, business appointment schedules, all meticulously timed to cohere with the new clock units of hours, minutes, and seconds. Everything was planned out in advance and assigned an appropriate time and duration. This included work schedules.

As early as 1700, companies such as the Crowley Iron Works in England were establishing formal work schedules for their employees. At the Crowley factory the owner designed a detailed code, which ran over one hundred thousand words, to oversee the employee's schedule.[9] Concerned over employee laxity, Crowley ordered the monitor and warden of the mill to keep a daily time sheet for each employee "entered to the minute."

> Every morning at 5 o'clock, the warden is to ring the bell for beginning to work, at eight o'clock for breakfast, at half an hour after for work again, at twelve o' clock for dinner, at one to work and at eight to ring for leaving work and all to be lock'd up.[10]

Every Thursday, the warden delivered the time sheet of each of the workers to the owner, along with the following affidavit:

> This account of time is done without favour or affection, ill-will or hatred and I do really believe the persons above mentioned have worked in the service of John Crowley Esq the hours above charged.[11]

A hundred years later, factory owners installed the first time-recording clocks to monitor automatically the exact arrival and departure times of employees. The first such device was a long case clock with a rotating dial and pegs located at every quarter-hour mark. The worker would pull a cord which pushed in a peg and recorded his time in or out. If the worker was one minute off in coming or going, he would be penalized a quarter-hour.

To encourage workers to adapt to the new clock consciousness and work schedules, employers combined harsh penalties and discipline with incentives and rewards. Fines were levied for showing up late or for "slothing" on the job. When the labor supply fell short, owners were forced to switch over to incentives to cajole workers into accepting the new time discipline. Josiah Wedgwood, Charles Darwin's father-in-law and a well-respected factory owner in England in the late eighteenth century, ordered his clerks to show special favoritism to punctual workers:

> Encourage those who come regularly to their time, letting them know that their regularity is properly noticed, and distinguishing them by repeated marks of approbation, from the less orderly part of the work-people, by presents or other marks suitable to their ages.[12]

The work schedule was far more demanding than any other time invention in history. Every moment of the worker's time was subject to control from above. Even talking between employees was restricted, or forbidden altogether, lest it interfere with the preset time schedule. In John Marshall's flax mills, the following edict was issued in 1821:

> If an overseer of a room be found talking to any person in the mill during working hours, he is dismissed immediately.[13]

In the new scheme of things, "time was money" and any loss of time was to be avoided at all costs.

The work schedule often came into direct confrontation with that older form of social control, the calendar. People had been conditioned to the Christian calendar for nearly fifteen hundred years. It represented the dominant social and political force in their temporal life. Factory owners and merchants attempted to deflate the importance of the calendar but were confronted by worker resistance each step of the way.

As we learned earlier, the Christian calendar was peppered with holidays, saints' days, and feast days. With time now a precious commodity, factory owners were beside themselves over how to get around ecclesiastical time and secure worker acceptance of the more demanding work schedule. In South Wales, according to economist Sidney Pollard, workers were still losing "one week in five" as late as the 1840s as they took time off to celebrate various calendrical rituals and events.[14]

Throughout the long struggle between the bourgeois class and the workers over the new time orientation, the clock remained at the center of the dispute. Employees accused the owners of "fixing" the clocks to cheat the workers out of their wages. In one account, a factory worker testified that:

> The clocks at the factories were often put forward in the morning and back at night, and instead of being instruments for the measurement of time, they were used as cloaks for cheatery and oppression. Though this was known amongst the hands, all were afraid to speak and a workman then was afraid to carry a watch, as it was no uncommon event to dismiss anyone who presumed to know so much about the science of horology.[15]

For the most part, the new class of owners was unsuccessful in converting farmers and tradesmen into disciplined factory workers. They were too settled into the temporal orthodoxy of an earlier epoch. But it soon became apparent that their children, still temporally unformed, provided a much more convenient labor pool for the new industrial technology. Child labor was cheap and could be easily molded to the temporal demands of the clock and the work schedule. By spiriting children away at the tender age of five to seven to work up to sixteen hours a day inside dimly lit and poorly ventilated factories, the owners insured themselves a captive and manipulable work force that could be thoroughly indoctrinated into the new time frame.

At one time or another, up to one-third of the factory work force was comprised of children ranging in age from five to eighteen.[16] Children were often beaten into submission by foremen.

In the more progressive factories, beatings gave over to verbal abuse and psychological humiliation. In John Wood's spinning factory, a child found guilty of violating the rules of the "work schedule" would have to walk up and down the room holding up a card listing his offense. If guilty of a particularly serious indiscretion, the child would be forced to confess his crime to his fellow workers.[17] In the Witts and Rodick silk mill in Essex, children who violated the rules of the work schedule had to wear "degrading dress."[18]

Factory owners also used the incentive system to condition their child labor to the work schedule and to increase their productivity. In the silk mills, Pollard reports, clothes were often given away as prizes for exemplary work. In one factory, the hardest-working girls received dolls and the best boys received a slab of bacon and threescore of potatoes.[19]

Getting workers to accept the new conception of time ultimately depended on the ability of the owners to convince the average laborer that through a combination of diligence, punctuality, discipline, and hard work he could better his lot in life, secure greater material wealth, improve his status in society, and assure a better future for his children. In short, the worker needed to be made ambitious. He had to be liberated from the weary wheel of history, the ageless cycle of village life in which generation followed generation in endless repetitions of familiar and time-honored tasks.

The bourgeois temporal orientation was linear, not cyclical. The bourgeoisie had emancipated the future from nature and the gods and made it a secular frontier to be tamed by the exercise of human will and determination. Now they were calling upon the workers to accept their radical new time values. They received ample cooperation in this mighty effort from the Protestant evangelical ministers. The clergy attacked every element of the worker's life that might in any way impede his proper performance on the job. His insobriety, his foul language, his laxity, his lack of ambition were all made the object of countless sermons and public speeches. As E. P. Thompson wryly observed in his essay, "Time, Work-Discipline, and Industrial Capitalism": "Long before the pocket watch had come within the reach of the artisan," the evangelical clergy was "offering to each man his own interior moral time-piece."[20] One of the leading evangelical voices of late-seventeenth-century London, R. Baxter, wrote in his *Christian Directory*, "A wise and skilled Christian should bring his matters into such order, that every ordinary duty should know his place, and all should be . . . as the parts of a clock or other engine, which must be all conjunct, and each right placed."[21]

The church pulpit was not the only forum given over to spreading the good news about the new bourgeois time concept. The educational system also took up the challenge of preparing the next generation in the ways of the new temporal orientation. Writing in 1772, one social commentator declared that the schoolroom should be a training ground in the "habit of industry" and that at the earliest age every child should become "habituated, not to say naturalized to Labor and Fatigue."[22]

Educators enthusiastically embraced the new concept of scheduling and were quick to transpose the disciplined rhythms of factory work directly into the schoolroom. The new time rules governing the schooling of children have remained virtually untouched to the modern day. Here is a description of the school schedule adopted in the early 1800s:

> The superintendent shall again ring—when, on a motion of his hand, the whole school rise at once from their seats;—on a second motion, the scholars turn;—on a third, slowly and silently move to the place appointed to repeat their lessons— he then pronounces the word "Begin."[23]

In schools, punctuality and strict adherence to time schedules became as important, if not more important than reading, writing, and arithmetic. A French minister of education, reflecting on the superiority of the French system of schooling, was led to boast that "he could consult his watch at any moment of the day and say whether every child in France, of a given age, would be doing long division, reading Corneille, or conjugating Latin verbs."[24]

Inside the schoolroom, astride the church dais, on the factory floor, the new urban culture was being entrained to a new temporal catechism. The clock and the schedule were being indelibly imprinted into the consciousness of the culture. To be modern was to be punctual, disciplined, fast-paced, and future-directed. Spontaneity, irregularity, laxity, and the unhurried ease that accompanied a less materialist, more traditional medieval culture was being abandoned in favor of a restless, driving Promethean vision. The communal spirit of the manor estate and the quiet country village was being replaced by the atomistic will of city life. The new man and woman were taught to expose the totality of their lives to an exacting schedule and to fill in every moment with a productive task. The clock culture called forth a new faith: the future could be secured if everyone would only learn to be on time. Securing the future in this brave new world depended less and less on good works and God's grace and increasingly on being punctual.

The reward for a lifetime of service in the medieval era was the administering of final absolution. In the new clock culture, a lifetime of service was honored by the bestowal of an engraved gold watch for fifty years of unswerving punctuality. In the new era, being "a good company man" replaced the idea of being "a faithful servant of the Lord." Time was stripped of its sacred context and made into a utility to advance the productive goals of an increasingly secular civilization.

7

PROGRAMS
AND COMPUTERS

Today, as we make the transition into the postindustrial age, the computer program is beginning to join the clock schedule as a new time-allocating device. While both the schedule and the computer program are temporal tools for planning out future events, only the program is capable of predetermining the "exact" way the future is to unfold. The schedule determines how minute segments of the future will be used and provides detailed instructions governing sequence, duration, tempo, and coordination of activities. But while schedules are instructions for the future, they do not rigidly govern how that future will develop. Between the instructions and the future events are people, who must mediate the plan and the execution of the plan. This being the case, there is never an exact correspondence between the schedule's conception of how the future is to be filled in and how the future actually is filled in. As rigid as the schedule is, it cannot be used to absolutely control future segments of time because it must always contend with human modification, error, and caprice. As long as human beings remain the agents of change, future events can never be totally controlled by preestablished schedules.

But computer programs are more than instructions. Programs not only plan out the future but are capable of executing it as well. Programs can be designed to determine in advance the sequence, duration, and tempo of an event. They can determine when in the future an event will unfold and they can coordinate and synchronize the interaction of multiple events over time. With programs, it is possible to lock future events into a predetermined course in a way schedules cannot. Programs are now in place that require no human intervention whatsoever in the process of filling in the future. A sophisticated computer program can instruct automated machinery to make a product, or deliver a service, without ever having to rely on human participation in the unfolding of the activity.

Deep inside a pine forest near Mount Fuji in Japan is a row of very plain-looking yellow buildings. Inside the dimly lit structures sophisticated mechanical robots work nonstop twenty-four hours a day making parts for machine tools. The plant occupies 54,000 square feet and is supervised by a single human being whose only job is to monitor the machines on closed-circuit television. Computer programs run the entire operation. They determine what is to be done and how long it is to take. The programs coordinate each one of the myriad activities of the plant without any human beings present on the factory floor. The only human participation in the process is in the design of the program. The long-term psychic and social repercussions of this temporal shift are earthshaking.[1]

No human being actually participates in the unfolding of the future at the Japanese factory. Even the programmer never takes part in the future he or she has designed. With computer programs, society begins to relate to the future in an entirely new way. The programmer participates in the future only in the instructional preparation of it. The rest of society is even less involved. They become voyeurs, watching various predetermined futures unfold with little ability to affect or change activities that have already been imprinted into the programs well in advance. Large sectors of society are forced to experience future segments of time and future events without being able to participate directly and help shape their outcome.

While an increasing number of programs are designed to eliminate human participation in the unfolding of the future, they also tend to remove the individual user from subjective past experiences. In this respect, the program differs once again from the way the schedule organizes time. With the schedule, each person brings his or her own past into the future. Personal past experiences are used as both a resource and a guide to future actions. It is true, of course, that script, print, and film have all provided human beings with external memory sources. But in each case, the individual integrates these other stored experiences with subjective memory to make decisions about how to act on the future. Even on the factory floor, where preestablished schedules detail how the worker is to fill in his future, the employee still relies on his own past experience to modify or amplify the instructions to be carried out.

Computer programs undermine subjective memory. Users rely less and less on their own personal memories of past experiences and events as a guide to future actions and become more reliant on the data stored in the program's memory bank. Even in the modification of existing programs or in the writing of new programs, personal memory becomes more and more limited to the task of data recall, remembering specific codes for gaining access to stored information.

Computer programs, then, eliminate a great deal of active involvement in one's past as well as in one's future. The kind of past that is available to draw upon is largely determined by the available software and data stored in the memory bank and the selective ways the programs are designed to use that data. The future is largely predetermined by the way the program prints out its instructions. In the factory at the base of Mount Fuji, where the only human being on the premises is the monitor, personal initiative plays a negligible role in determining how the future is to unfold on the factory floor. If a warning light comes on, he can push a button or plug in a coded instruction rerouting the activity. Yet even these maneuvers have been preset, coded into the software. The program, after all, determines if and when the warning light goes on and the program instructs the monitor in the proper procedures to follow in case of emergency.[2]

As an instrument of social control, the program far exceeds the power of either the calendar or the schedule in its ability to dictate the terms of the past and impose its will over segments of the future. By effectively programming more and more of the activities of society, those in power will be able to increasingly separate citizens from personal involvement in the decisions that affect their lives. Computer programs introduce a new level of determinism into the social process. By automating the unfolding of future events, computer programs leave the individual a passive victim, forced to live within the narrow confines of preprogrammed scenarios laid out for him.

The power of the program flows directly from the computer technology that gives rise to it. As a timepiece, the computer differs significantly from the traditional mechanical clock. The first thing to recognize is that the computer operates by electronic current, not by gears. Electricity moves near the speed of light. As Marshall McLuhan pointed out in *Understanding Media*, the electric current shortens duration to the point of near simultaneity.[3] This shortening of perceived durations changes our entire consciousness of time. With the clock we think of time as ticking. With the new computer timepiece, we begin to think of time as pulsing.

The second distinguishing feature of the computer is its temporal creativity. David Bolter, author of *Turing's Man*, points out that while clocks are all set to the same exacting sequence, duration, and rhythm, the computer is free to manipulate all three of these temporal dimensions by merely changing the program.

The computer imprints a unique temporality into every program. Every computer has an electronic timer in its central processor. The timer releases electrical impulses at specific intervals allowing the central processor "to execute one by one the instructions given in its program."[4]

> The electronic timer provides the measure by which the pro-
> cessor ticks its way through its calculations, ensuring that the
> electrons have settled down, that one step is finished before
> the next is begun. The instructions themselves may require
> varying amounts of time. . . . This variation must be taken into
> account by the sequencing mechanism, which decides how
> many pulses of time to allot to each instruction.[5]

Bolter argues that time is a resource for the computer just as
coal is a resource for the steam engine. Time is used to transform
"billions of countless impulses of electrical energy into useful
instructions for manipulating data."[6] The difference, then, between clocks and computers is that "an ordinary clock produces
only a series of identical seconds, minutes, and hours; a computer
transforms seconds or microseconds or nanoseconds into information."[7] With this new timepiece, time is no longer a single
fixed reference point that exists external to events. Time is now
"information" and is choreographed directly into the programs
by the central processor. With computers we enter the age of
"multiple times."[8] Every program has its own sequences, durations, rhythms, its own unique time.

While the clock established the notion of artificial time segments—hours, minutes, and seconds—it remained tied to the
circadian rhythm. The clock dial is an analogue of the solar day,
an acknowledgment that we perceive time revolving in a circle,
corresponding to the rotation of the earth. In contrast, computer
time is independent of nature: it creates its own context. A digital
timepiece displays numbers in a vacuum—time unbound to a
circadian reference. Computer time, then, is a mathematical abstraction that attempts to separate us from the pulls and periodicities of the natural world.

8

THE EFFICIENT SOCIETY

Clocks and schedules, and computers and programs, have transformed the sociology of human existence. The modern time world is fast-paced, future-directed, and rigorously planned. The new time technologies have changed our way of life and, in the process, have effected a fundamental change in the value orientation of Western culture. The artificial time worlds we have constructed have been accompanied by a radical new temporal value: efficiency. With its introduction, the modern temporal orientation is complete. Efficiency is both a value and a method. As a value, efficiency becomes the social norm for how all human time should be used. As a method, efficiency becomes the best way to use time to advance the goal of material progress.

To be efficient is to minimize the time in which a task is completed or a product produced and to maximize the yield, expending the minimum amount of energy, labor, or capital in the process. In less than two hundred years, efficiency has risen from obscurity to become the overriding value of society and the primary method for organizing the activities of the human family. Efficiency is the hallmark and trademark of contemporary culture. It binds the various temporal features of the modern world into a single unifying focus. Today efficiency pervades every facet of life: it is the primary way we organize our time and has burrowed its way into our economic life, our social and cultural life, and even our personal and religious life.

We have institutionalized efficiency through the schedule and now the program. Every activity is scheduled or programmed in advance so that we may use time in the most efficient manner possible. Optimizing schedules and programs means optimizing efficiency.

Efficiency was introduced into the popular culture through the workplace. If the first task of industrial capitalism was to make the workers punctual and to discipline them to accept clock time, the next major task was to make them efficient.

Efficiency is a product of three major economic innovations, each of which radically transformed people's relationships to their tools and to their fellow beings: division of labor, mass production, and the principles of scientific management. These represent the cornerstones of the industrial pyramid, and each has played a key role in making efficiency the overriding temporal conception of the industrial way of life.

Efficiency's ascent to power began with the introduction of division of labor. Economic historian Harry Braverman contends that "in one form or another, the division of labor has remained the fundamental principle of industrial organization."[1] The first philosopher to articulate the importance of division of labor in industrial production was Adam Smith. Writing in *The Wealth of Nations*, Smith contended that the new principle of division of labor provided a means of "saving time" in the production process:

> This great increase in the quantity of work, which, in con-
> sequence of the division of labor, the same number of people
> are capable of performing, is owing to three different circum-
> stances; first, to the increase of dexterity in every particular
> workman; secondly, to the saving of time which is commonly
> lost in passing from one species of work to another; and lastly,
> to the invention of a great number of machines which facilitate
> and abridge labor, and enable one man to do the work of many.[2]

Adam Smith came to this realization by observing the great strides taking place in watch manufacturing. It was there that modern industry first began to apply the principles of division of labor to increase production. As far back as 1703, master clock- and watchmaker Thomas Tompion was mass-producing time- pieces. His biographer, Symonds, says that Tompion's success lay in organizing his workshop "in a way hitherto unknown in the English handicrafts."[3] Sir William Petty, one of the distin- guished political economists of the period, wrote the following description of the new method Tompion and others were applying to production:

> In the making of a watch, if one man shall make the Wheels, another the Spring, another shall engrave the Dial-Plate, and another shall make the Cases, then the watch will be better and cheaper, than if the whole work be put upon any one Man.[4]

Division of labor meant that more goods could be produced in "less time" at a cheaper cost per unit.

The division-of-labor concept was followed in close order by the second major economic innovation, the introduction of mass production principles. Eli Whitney introduced the idea of mass production in 1799. Frustrated over the long time delays that resulted from having to teach workers the necessary skills to make the various component parts that went into the assembly of a finished product, Whitney developed the idea of mass-producing standardized interchangeable parts that could be easily assem- bled by unskilled laborers. He applied the new principles of mass production to the making of muskets.

> To eliminate guesswork by eye, he invented jigs, or guides for tools, so that the outline of the product would not be marred by the fallibility of a shaky hand or imperfect vision. He made automatic stops that would disconnect the tool at the precise

depth of diameter of a cut. He made clamps to hold the metal while the guided chisels or milling wheels cut it. By dividing his factory into departments—one for barrels, one for stocks, one for each lock piece—the parts could be brought into an assembly room and put together in one continuous uninterrupted process.[5]

Whitney's new mass production process became known as the "American Method." Its principles soon spread to the watch industry, where they were further refined and eventually served as a model for the rest of American industry. The man responsible for applying Whitney's idea to watch production was Aaron L. Dennison. He joined forces with Whitney to set up a company which later became known as the Waltham Watch Company, the first mass production watch company in the United States.[6]

The principles of division of labor and mass production were both intended to save time. To be effective, they required the setting up of detailed work schedules so that every operation would be subjected to rigorous time standards. To ensure that every moment of the production process would be used to maximize output, a third and final innovation was introduced into the industrial process. It was called scientific management and its author was Frederick W. Taylor.

Taylor made efficiency the *modus operandi* of American industry and the cardinal virtue of American culture. His work principles have been transported to every sector of the globe and have been responsible for converting much of the world's population to the modern time frame. He has probably had a greater effect on the private and public lives of the men and women of the twentieth century than any other single individual. Economic historian Daniel Bell says of Taylor:

If any social upheaval can ever be attributed to one man, the logic of efficiency as a mode of life is due to Taylor. . . . With

scientific management, as formulated by Taylor in 1895, we
pass far beyond the old, rough computations of the division of
labor and move into the division of time itself.[7]

Taylor's principles of scientific management were designed
with one goal in mind: to make each worker more efficient. His
primary tool was the stopwatch. Taylor divided each worker's task
into the smallest, visibly identifiable operational components, then
timed each to ascertain the best time attainable under optimal
performance conditions. His time studies calibrated worker performance to fractions of a second. By studying the mean times
and best times achieved in each component of the worker's job,
Taylor could make recommendations on how to change the most
minute aspects of worker performance in order to save precious
seconds, and even milliseconds, of time. Scientific management,
says Harry Braverman, "is the organized study of work, the analysis of work into its simplest elements and the systematic improvement of the worker's performance of each of these elements."[8]
Taylor considered his work principles to be scientific to the
extent that he was able to eliminate all nonquantifiable elements
of worker behavior. His time studies reduced every aspect of work
to the dictates of time. Worker performance could now be reduced
to numbers and statistical averages that could be computed and
analyzed to better predict future performance and to gain greater
control over the work process itself.
Taylor relied on a new approach to management. The stopwatch and statistics ruled the factory floor. "Management," by
the way, seemed an altogether appropriate term to affix to the
new scientism. Braverman reminds us that "manage" comes from
the Latin *manus*, which meant "to train a horse in his paces, to
cause him to do exercises of the manege."[9]
Taylor believed that the best way to optimize the efficiency of
each worker was to assert complete control over all six temporal
dimensions: sequence, duration, schedule, rhythm, synchronization, and time perspective. No aspect of the worker's time was

to be left to chance or to worker discretion; from now on, the worker's time would fall under the absolute control of management. The most efficient state, said Taylor, was the most autocratic. Taylor's principles of scientific management represented the ultimate politicization of the new industrial time. Braverman argues that Taylor's work "may well be the most powerful as well as the most lasting contribution America has made to Western thought since the Federalist Papers."[10]

Taylor's first principle of scientific management was for management to seize control over the knowledge of the work process that had previously been in the hands of the workers. From now on, Taylor stated:

> The managers assume . . . the burden of gathering together all of the traditional knowledge which in the past has been possessed by the workmen and thereof classifying, tabulating, and reducing this knowledge to rules, laws, and formulae.[11]

Taylor's intention was to sever the labor process from the skills of the workers. Those skills were to reside only in the hands of management.

Taylor's second principle flowed directly from the first. Having gained a monopoly over the knowledge required to do the work, management must then assume the authority to plan and direct the work on the shop floor. Denied firsthand knowledge of how their work was to be done, the workers would become totally dependent on management in the execution of their tasks.

Taylor believed that as long as the workers maintained both knowledge and control over how their work was to be done, it would be impossible to elicit maximum efficiency. Left on their own, workers would let other "human" considerations enter into the work process. Feelings and emotions would come to the fore, tempering and even undermining the prospect of attaining maximum efficiency. For example, workers might consciously choose to moderate their work pace to accommodate the needs of slower employees. They may even relax their concentration by occasional socializing. Taylor argued that "if the workers' execution is guided by their own conception, it is not possible . . . to enforce upon them either the methodological efficiency or the working pace desired by capital."[12]

In order to secure maximum efficiency in the execution of the work process, a third and final principle of scientific management was called for: the implementation of the "work schedule." It was here that management cemented its control over the total work time of each of its employees.

> The work of every workman is fully planned out by the management at least one day in advance, and each man receives in most cases complete written instructions, describing in detail the task which he is to accomplish, as well as the means to be used in doing the work. . . . This task specifies not only what is to be done, but how it is to be done and the exact time allowed for doing it. . . . Scientific management consists very largely in preparing for and carrying out these tasks.[13]

Taylor believed that the key to making a worker more efficient was to strip him of any capacity to make decisions regarding the conception and execution of his task. In the new scientifically managed factory, the worker's mind was severed from his body and handed over to the management. The worker became an automaton, no different from the machines he interacted with, his humanity left outside the factory gate. On the factory floor, he was an instrument in the production process, a tool whose performance could be timed and improved on with the same cool detachment and scientific rigor as might be applied to the machinery itself.

In the years following Taylor's pioneering efforts, the principles he first enunciated were further refined. New scientific tools allowed more exacting controls to be exercised over the work process. The most interesting advance in the principles of scientific management occurred with the introduction of motion-

and-time studies. This development was the brainchild of Frank B. Gilbreth, one of Taylor's early disciples.

Gilbreth filmed the movements of each worker in order to establish standard times for each body motion. Virtually every movement on the factory floor and in the clerical offices was analyzed and assigned an optimum time, usually calibrated down to a fraction of a second. The various movements, in turn, were assigned standardized names, using machine terminology. For example, "contact grasp" referred to picking an object up with fingertips. "Punch grasp" meant thumbs opposing finger. "Wrap grasp" meant wrapping one's hand around the object.

If the task required picking up a pencil, it would be described in the following manner: transport empty, punch grasp, and transport loaded. Each movement was assigned a standardized time. The sum total of the individual times associated with each movement would be the standard time for completing the task. The time calibration in the Gilbreth motion-and-time study was perfected down to ten-thousandths of a minute.[14]

Today the science of motion-and-time studies is far more sophisticated than anything Gilbreth could have imagined. Sound waves are used to detect minute changes in body movement and are calibrated to an accuracy of .000066 minutes.[15] Even the worker's visual movements can now be timed and standardized. Through a process called electrooculography, it is possible to time every single shift in eye movement as the worker scans the various monitors and controls with which he or she is working. Standardized times for each shift in eye movement are calibrated, providing a norm for measuring the optimum efficiency of all eye movements.[16]

Motion-and-time studies have been used successfully in establishing time efficiencies in every work environment. In the clerical field, standardized times have been assigned to the smallest tasks, as evidenced by the following motion-and-time chart compiled by the Systems and Procedures Association of America.[17]

Open and close	Minutes
File drawer, open and close, no selection	.04
Folder, open or close flaps	.04
Desk drawer, open side of standard desk	.014
Open center drawer	.026
Close side	.015
Close center	.027
Chair activity	
Get up from chair	.033
Sit down in chair	.033
Turn in swivel chair	.009
Move in chair to adjoining desk or file (4 ft. maximum)	.050

Taylor and his disciples turned efficiency into a science. They inaugurated a new ethos. Efficiency was officially christened the dominant value of the contemporary age. From now on, no other consideration would be allowed to compete with or undermine this ultimate value. It would not be long before Taylor's principles would find their way into the rest of the culture, changing the way we lived and interacted with each other in the modern world. The new man and woman were to be objectified, quantified, and redefined in clockwork and mechanistic language. They were to be turned into a factor and then a cog in the production process. Their labor was to be divided, standardized, and regulated to a fraction of a second, then regimented to the task of achieving maximum material output. Above all, their life and their time would be made to conform to the regimen of the clock, the prerequisites of the schedule, and the dictates of efficiency.

The clock culture and the industrial system journeyed together into the future—two forces inseparably linked, each helping to

define and shape the dimensions of the other. Determined to
impose the new temporal value of efficiency and the new industrial mode of production onto the rest of the world, the American
and European powers, in consort with merchants, industrialists,
and financiers, began the task of creating a new standard world
time.

Up until the latter part of the nineteenth century, most of the
world still ran by local times. Each culture, each geographic
region, and each nationality had its own system of time reckoning
and time keeping. These local time-keeping systems were still
tied to traditional calendars and regulated by differing astronomical calculations, variations in seasonal and environmental phenomena, and local traditions. The great number of time systems
made the planning and coordinating of economic activity difficult,
if not impossible. What was needed was a universal time-reckoning system that could bring the whole world under a single
unified time frame. Advocates of a new temporal standard force-
fully argued that the adoption of one world time would greatly
facilitate the complex scheduling of economic activities around
the globe, foster greater efficiency, advance the cause of material
progress, and make the world a more secure and safer place to
live in.[18]

The initial pressure for a universal time system came from the
railroad companies. In 1870, a rail passenger traveling from
Washington to San Francisco would have to reset his watch over
two hundred times to stay current with all the local time systems
along the route.[19] Twenty-three years earlier, the British railroad
companies had adopted a single national time.[20] Soon other countries with rail travel adopted standard times. In the larger nations
like the U.S., zones were established that divided the country
into separate time corridors. By the late nineteenth century, nations were pressing for a single world time. Naturally the British
favored using Greenwich, near London, as the place to locate
zero longitude for a world time-reckoning system. Precedent was
on their side, as navigators had long used Greenwich time as

their point of reference. The French were opposed and suggested that the Paris Observatory would be a more suitable point of reference to regulate the new universal time system.[21]

In October 1884, the International Meridian Conference voted to make Greenwich timekeeper for the planet.[22] The French continued to hold out for another twenty-seven years but finally surrendered to international pressure and accepted Greenwich as the zero meridian. In 1912, Paris hosted the International Conference on Time for the industrial countries of the world and made it official. From that time on, the earth and all of its inhabitants would be subjected to one world time-reckoning system.

The standardization of world time marked the final victory for efficiency. Local times had long been tied to traditional values, to nature, to the gods, to the mythic past. The new world time was bound only to abstract numbers. It flowed evenly and remained aloof and detached from parochial interests. The new time expressed only a single dimension: utility. The industrial nations adopted a universal standardized time frame simply because it was more efficient. The world was being readied for the new temporal imperialism.

It took six hundred years to revolutionize the temporal orientation of Europe. It took only one-third that time to extend the temporal revolution to countries and cultures across the globe. In the sixteenth, seventeenth, and eighteenth centuries, European armies colonized the territories of the planet. In the nineteenth and twentieth centuries, European and American industry colonized the time frame of much of the rest of the world.

Although little has been written on the subject, there is no doubt that time has been a critical factor in redirecting entire cultures to the modern frame of mind. Industrialists, merchants, and traders have all been confronted by traditional time orientations that are incompatible with the temporal notions that make up the industrial way of life. Developing an industrial labor force and industrial markets in many countries has proven to be as formidable as it was in Europe, for many of the same reasons. Time awareness in traditional cultures is not cued to the clock, the schedule, and the value of efficiency. Economic development specialists are often highly critical of the time values of local cultures and are quick to disparage the indigenous labor force in terms reminiscent of the epithets leveled at European workers by bourgeois factory owners and merchants.

The new class of workers in certain cultures is said to be slothful, unpredictable, and unreliable. It is argued that they have no sense of the future and how to plan ahead. They are unpunctual and undisciplined. They care only for the moment. They are slow to initiate and slow to react. They are settled into traditional behavior patterns and unwilling to learn more efficient ways of doing things. In short, they do not share the time frame of the Western world.

Of course, there are exceptions. Some traditional cultures have been able to adjust to the new time frame with relative ease. Already destitute, they willingly accept the imposition of new conditions of time just as the Irish farmers of an earlier century did. Other cultures not only adapt to the new time demands but are able to master them, posing a direct commercial and economic threat to the Western powers. Whether a culture resists, surrenders to, or masters the new time orientation depends largely on its temporal predispositions.

As it has made its way around the world, the temporal imperialism of the modern age has elicited many different political responses. Nowhere have the responses been more strikingly different than in China and Japan. Both countries were subjected to Western attempts at colonization. China eventually surrendered and collapsed in the face of the colonial onslaught, allowing the European nations and corporate interests to carve up its geography into separate spheres of influence under the control of different flags. Japan, a tiny nation with far fewer people and resources, struck a different bargain with the West. It was never successfully colonized, and, in the end, it expropriated the Western forms of power to become a strong rival on the world scene.

To understand why these two cultures responded so differently to Western imperialism, it is necessary to examine their temporal orientations. Their concepts of time played an important, if not critical, role in shaping their reaction to the industrial and political armies of the West.

According to Robert H. Lauer, who has made a detailed study of the time orientation of the two countries, China did not "confront time with the Western obsession to maximize activity in a minimum of time, and therefore was never willing to fight for that which could be obtained in other ways."[24] In China the tempo of change was, historically, slow. Then, too, the Chinese viewed time as cyclical rather than linear. All of the great Chinese religions—Taoism, Confucianism, and Buddhism—taught that time and history endlessly repeat themselves in strict obedience to planetary movements. The notion of progress was absent from the Chinese time frame. While the industrial West had come to think of the future as the place where the earthly Eden will unfold, the Chinese had long believed that the golden age had already passed from view.[25]

The West paid homage to the future, the Chinese to the past. Consequently, when new issues arose, the Chinese resolved them by referring to precedents, tradition, or custom. According to Lauer, "intellectual controversies were waged on the basis of conformity or deviation from the past, which had unquestionable authority."[26] William Parsons observed that "no dictator ever ruled with greater power than 'Precedent in China' and that the Chinese tend to regard the future as merely an opportunity to relive the past."[27]

The Chinese, then, perceived time as slow-moving, cyclical, predetermined, and saw themselves as guardians of past glories rather than initiators of new visions. The result, says Lauer, was that the Chinese

> foundered because of the limited number of responses allowed
> by that temporality. As long as China existed in a temporality
> that was cyclical, predetermined, and bound to a particular

vision of the past, she remained stagnant and helpless before the West.[28]

The Japanese time orientation was radically different. The Japanese viewed change as a rapid and accelerating force, and time as more linear than cyclical. Dogen, the thirteenth-century philosopher who introduced the Soto Zen sect to Japan, viewed reality as fluid and impermanent. The Shinto religion also stressed the idea of time as a linear process and the future as a place to make history.[29]

The Japanese reverence for tradition was also quite different from that of the Chinese. Instead of idealizing the past, the Japanese preferred to think of traditions in a more instrumental manner. Sinologist Nyozekan Hasegawa argues that the importance of tradition "lies not so much in the preservation of the cultural properties of the past in their original form as in giving shape to contemporary culture; not in the retention of things as they were, but in the way certain national qualities inherent in them live on in the contemporary culture."[30] The Japanese are comfortable with the idea of change. Not surprisingly, they are future-oriented rather than past-oriented. Further, the Japanese view time as a scarce resource, as does the West. In the Tokugawa era, extending from the seventeenth to the mid-nineteenth century, it was not uncommon for the government to encourage its subjects not to "waste time."[31]

All in all, the Japanese temporal orientation was more expedient than that of the Chinese, more pragmatic and instrumental, and more poised to constructing the future rather than protecting the past.

When the Japanese, therefore, were confronted with a challenge similar to that imposed upon China, their response was quite different. Unlike the Chinese, the Japanese did not view the future as predetermined, nor as requiring conformity to the past. The Japanese attempted to create their future rather than merely adapt it.[32]

China, Japan, and much of the world have fallen victim to the efficiency time frame of the West. Over the past two centuries the clock and the schedule have increasingly dominated the economic and social life of nations across the globe. Together they have radically transformed the cultural patterns of much of the world community. Wherever the industrial way of life has gained a foothold, people have been forced to turn away from their long-standing rhythmic relationship to nature and made to conform with the single-minded value of increased efficiency.

Now with the clock and the schedule joined by an even more powerful time technology, the computer and the program, efficiency has assumed an unchallenged position in the social scheme of things, becoming the premier value of our age.

All around the world, businesses are making the transition to computer-run operations. As they do, the flow of work-related activity accelerates dramatically, and efficiency becomes reduced to a nanosecond time frame. Work that often took considerable time to "organize" and "produce" is now expected to be "programmed" and "processed" in a fraction of the time. Where a secretary used to average about 30,000 keystrokes per hour, the average VDT operator is expected to perform 80,000 keystrokes in the same amount of time. In brokerage firms, personnel responsible for customer accounts are now expected to handle one call every minute and a half, providing the shareholders with complete stock market information accessed from the computer console at their work station. An architect using advanced computer-aided design (CAD) can now "make nineteen times more decisions per hour than does a pencil-wielding colleague."[33] Mike Cooley, past president of the Designers Union of Great Britain, and an expert on computer-aided design, says that there are systems currently in use "where the decision-making rate of the designer can be forced up by 1800 percent." The frantic tempo established by these systems puts enormous pressure on the designer, often crippling his or her creative contribution. According to Cooley, "the rate at which the computer demands decisions reduces his creativity by 30 percent in the first hour, 80 percent in the second hour, and thereafter he's just shattered."[34]

To insure "optimum interface" between the machine and the operator, computer experts are beginning to map the "peak performance age" for people working with computers in various fields of knowledge. People of different ages are scored on their ability to respond quickly to various problems on the visual display unit. The speed of their response is then used to calculate peak performance age. The peak performance age of a mathematician is twenty-three. A theoretical physicist peaks at twenty-seven and a structural engineer at thirty-four.[35]

The computer represents the ultimate technological expression of our culture's obsession with efficiency. It was introduced into our life as a means of helping us make the most efficient use of time. In the computer realm, time is information and information is utility. Time used in ways that cannot be broken down, quantified, and measured in terms of efficient inputs and outputs is of little or no value. Craig Brod's account of one of his patients, a supermarket cashier, is indicative of the way the new computer world transforms time into speedy, useful information while eliminating temporal activity without immediate instrumental value. When Alice's employer installed electronic cash registers, built into the computer-run machine was a counter that "transmits to a central terminal a running account of how many items each cashier has rung up that day."[36] Alice finds that she no longer "takes the time" to talk with customers as it slows down the number of items she can scan across the electronic grid and might jeopardize her job.[37]

In Kansas a repair service company keeps a complete computer tally of the number of phone calls its workers handle and the amount of information collected with each call. Says one disgruntled employee, "If you get a call from a friendly person who wants to chat, you have to hurry the caller off because it would count against you. It makes my job very unpleasant."[38]

In an effort to speed up the processing of information, some visual display units are now being programmed so that if the operator does not respond to the data on the screen within seventeen seconds, it disappears. Medical researchers report that operators exhibit increasing stress as the time approaches for the image to disappear on the screen: "From the eleventh second they begin to perspire, then the heart rate goes up. Consequently they experience enormous fatigue."[39]

The computer is even being used to speed up normal conversation. Companies like Sony and Panasonic are now marketing variable-speech-control cassette tape recorders equipped with a specially designed, solid-state speech compression chip. The variable-speech-control chip speeds up the tape-playback motor while clipping off tiny audio fragments—lapping about 10 milliseconds off each sound. The remaining sounds are electronically stretched, producing a fast-paced, intelligible narrative.[40]

With the compression chip, it is possible to listen to any voice cassette in half the time without being subjected to the high-pitched sounds that result in fast-winding the traditional tape recorder. With variable speech control, a sixty-minute cassette can be listened to in half an hour. According to industry sources, about a million people now "speed listen," and millions more will in the years to come as the compression chip makes its way into schoolrooms, offices, and homes across the country.

The far-reaching implications of this accelerated time frame are difficult to assess. Like the clocks and schedules that preceded them, computers and programs cut deep into humanity's relationship with nature, severing many of the remaining temporal bonds between our species and the larger environment. Computer time bears no relationship to the rhythms of nature. It is an arbitrary temporal marker willed into existence by human ingenuity. The computer reduces time to numbers and turns duration into uniform segments that can be added, subtracted, accumulated, and exchanged. While the computer turns time into a purely manipulable commodity, programs turn human beings into instruments to serve the new efficiency time frame. With the computer and program, each person's immediate future can be predetermined down to the tiniest artificial time segments of milliseconds and nanoseconds. Computers and programs represent a new form of social control, more powerful than any previous means used to marshal and regiment human energy.

PART III

THE
POLITICS
OF
PARADISE

9

THE TIMELESS STATE

As we have just seen, social order in every society is maintained with time-allocating devices like calendars, schedules, and programs. Those in power are able to convince the people to accept the time restraints imposed on them by promising that in return for their time sacrifices, they will be rewarded in the future with entrance into an idyllic, "timeless" paradise.

Most societies create an "image of the future," an ideal state that serves as a goal and a goad, an incentive and a prod toward which to strive. These images of the future incorporate the dreams and visions, hopes and aspirations, of the collectivity. The state is the caretaker of the communal vision. Effective rule, in every society, depends on the ability of those in power to establish a compelling image of the future and then convince the people to sacrifice their time in hope of gaining access to the perfect kingdom that exists just beyond the time horizon.

If there is a common denominator, a baseline proposition that permeates much of Western culture, it is the desire for something "different and better."[1] That yearning becomes incorporated into images of the future. Societies imagine not just a better future, but a perfect future. They construct an ideal world somewhere over the temporal horizon and then act in consort to perfect the vision they have laid out.

In each case, the ideal world envisioned exists in a "timeless" realm freed from the constraints of historical duration. Western images of the future are, for the most part, exuberant, fecund, and overflowing. We imagine a world at once ecstatic and orgiastic, complete and harmonious. Our images of the future incorporate our greatest hopes even as they reveal our innermost fears. We create them because we need to assure ourselves that all is not for naught, that death is not the end, that another world exists beyond our limited sojourn, a world where all our expectations will be met and all our doubts resolved. Every image of the future is a projection of humanity's insatiable drive for immortality.

Western culture has institutionalized its images of the future by way of religion and politics. Gods and great leaders are looked to as our protectors against the ravages of time. In his seminal work, *History, Time and Deity*, S. G. F. Brandon researched the great religions in world history and concluded that "the many and diverse beliefs and practices . . . are . . . found to have a common underlying motive—the defeat or avoidance of Time's inevitable process of decay and death."[2] Human beings have created religious images of the future in part as a refuge against the ultimate finality of earthly existence. Every religion holds forth the prospect of either defeating time, escaping time, overcoming time, reissuing time, or denying time altogether. We use our religions as vehicles to enter the heavenly kingdom, the netherworld, the promised land, or the state of nirvana. We come to believe in rebirth, resurrection, and reincarnation as ways of avoiding the inevitability of biological death.

Our spiritual quest for immortality has been accompanied by a secular quest. In the day-to-day world of politics, the term "security" has come to represent our insatiable drive to perpetuate ourselves at all costs. Hobbes, Locke, Rousseau, and other great political philosophers of the modern era have argued that the desire for security lies at the heart of the social contract.

Western man and woman have been obsessed with the notion of security. We want to make sure that the future can be made predictable and controllable. The state is supposed to look out for us, to protect us, to ensure our survival. It is our insurance policy against death.

The state maintains its hold over the body politic by continually invoking the image of an idyllic future kingdom, be it in heaven or here on earth. The masses have repeatedly shown their willingness to make sacrifices for the state in the hope of securing their future and guaranteeing their immortality.

Historian of philosophy Jean Guitton has said that all of our political experiments are like reconstituted Gardens of Eden. Every state projects a vision of an ideal world just beyond the horizon; a world devoid of conflict, injustice, and war; a world where happiness and contentment reign supreme. With every new governing order we journey back toward the courtyard of paradise, hoping to recapture a measure of the immortality we abandoned long ago. By creating the image of a secure, idyllic future, the state allows us to experience a small measure of immortality in the secular world.[3]

While Western societies have entertained images of the future as far back as the first recorded civilizations at Sumer and Mesopotamia, the first fully articulated future vision in Western culture is associated with the rise of the Jewish state in 2000 B.C. The Jews helped create the concept of history, thus ushering in one of the greatest advances in human consciousness since the beginning of time.

Earlier peoples had made little or no distinction between past, present, and future, preferring to experience reality as an ever-recurring state of existence. The cyclical sense of time mirrored the ecological and astronomical cycles, bonding human consciousness and culture to the rhythms of nature.

The Jews entertained a radically different idea about the nature of time. They replaced "Once upon a time" with a specific historic beginning: "the Creation."[4] They saw time as an irreversible plane in which specific, nonrecurring events took place. Abraham, Moses, and Joshua all existed at specific times and were associated with events that would never be repeated but that affected all things to come.

The Jews weakened the cyclical time frame of eternal return with the introduction of the linear time frame of history. In so doing, they began the process of separating human consciousness and culture from the periodicities of the natural world, creating the context for an ever-widening chasm between social time and environmental time in the centuries to come.

It should be pointed out that as long as time consciousness was cyclical, there was no need to create an image of a timeless future kingdom. In mythical cultures, duration was secondary to regeneration. Because everything repeated itself through endless cyclical regeneration, timelessness was built into the process. It was not necessary to project some timeless paradise at the end of history because there was no history to contend with. In traditional cyclical time cultures, the social order experienced a continual birth with each new ecological cycle.

Although cyclical time was partially challenged by the Jews with the emergence of a linear time frame, it was never totally defeated. Even today cyclical time consciousness continues to play a role in society, albeit greatly diminished. While we still regulate some of our social life by the changing seasons, the high technology–simulated environment we have created increasingly removes us from nature's periodicities.

The Jewish image of the future undermined cyclical time consciousness by singularizing the gods into an all-encompassing deity who was solely responsible for what transpired on earth and in the heavens. The Jewish God joined in a covenant relationship with his people, a pact, designed to open up the future to human will. The Jewish state agreed to give over to Yahweh its complete allegiance. God, in turn, agreed to make the Jews his chosen people. He would look over them, deliver them from bondage, help them prevail over their enemies, and, in the end, would lead them out of the long travail of human suffering and into a land bathed in milk and honey.[5]

The Jewish God was above history and also *in* history. This

God took part in the time of the world. While he resided in eternity, he was not above engaging in the day-to-day affairs that punctuated earthly life. Far from predictable, the Hebrew God intervened in specific events and engaged in dialogue with specific individuals whose names and activities were to be remembered as historical benchmarks.[6]

The Jews not only helped create the concept of history, but also imbued it with a mission. History was the place where God and his chosen people played out a relationship together. Above all else, it was a place where promises were made which would be fulfilled sometime in the future.[7]

The Jewish image of the future was one of hopeful expectation. In the beginning, God had created a perfect world. Adam and Eve corrupted his creation when they fell from grace. But in his mercy, God entered into an agreement with Abraham and his heirs that called upon them to use history as a forum to reconsecrate their relationship to him. God promised that if they used the future in a way that honored the covenant, he would bring history to an end and deliver his children into a new, more perfect world, the promised land.[8]

The Jews, then, introduced more than just history and purpose into the world; they introduced free will. Their image of the future depended on their own actions, as well as God's will. The future was no longer seen as fixed and predetermined but as a realm in which myriad possibilities existed. The future is where human beings make choices that affect their own destiny. The Jews were the first to conceive of the future as something they helped fill in.[9]

In Judaism, it was the great prophets and priests who controlled the collective image of the future. It was through them that Jehovah instructed his people. The prophets made predictions, forecast events, and most important of all, issued warnings and made prescriptions. As Frederick Polak observes:

> It is through the prophet that Jehovah, ever ready with advice, aid and admonition for his people, speaks. The prophet is God's

herald—more than a king, better than a priest. He is the way-shower both for rulers and ruled, and chief interpreter of the Covenant. Misfortune and divine wrath can just be averted in time by listening to the prophet's words.[10]

The prophets and priests served as the conscience of the community and the caretakers of the covenant. They admonished their people when they strayed from the image of the future that inspired the original pact with the Lord God, Jehovah. They were the overseers of kings and rulers, for they alone were privy to what the future held in store. The prophets and priests interpreted the future vision of the Hebraic people. They acted as judge and jury when that vision was desecrated and abandoned. In times of despair, they rekindled the ancient image of the future that first burned brightly in the eyes of the patriarchs Abraham, Isaac, and Jacob. The prophets and priests offered new hope. They predicted a better tomorrow, and told their people that the land of milk and honey beckoned them forth.

The prophets and priests were more than forecasters and in-spirationalists. They were also helmsmen. They attempted to steer the Jewish nation toward the future kingdom. They gave advice and counsel on how to proceed. They were shepherds who prodded their flock forward, over the trials and tribulations that history placed in front of them, mindful that their long odyssey would someday lead to the gates of an earthly paradise, fulfilling the promises made to their parents and their parents before them.[11]

The Jewish image of the future was followed in quick succession by the Christian worldview. The first Christians, the original apostles, identified with the Jewish image of the future and looked on Jesus as the long-awaited Messiah who had been sent by God to save his people.[12] Their hopes were fueled by an ever-worsening political crisis. Rome was continuing to tighten its stranglehold over Israel and it is in this context that the apostles came to see the historical Jesus as God's messenger, the holy one who would liberate the Jews from the yoke of oppression. Even after Jesus' death and resurrection, the disciples waited in anxious expectation of his immediate second coming, an event they associated with the final deliverance of the Jewish people, the end of history, and the beginning of the perfect kingdom.[13]

When Jesus failed to return as expected, it forced the great apostle Paul to recast the meaning and import of Christ's message in wholly new terms, setting the basis for a radical new image of the future.[14] Paul argued that it was not necessary to wait for the return of Jesus to experience salvation. Through the baptismal ritual and the Mass, each person could enter into a personal union with the Savior and be born again into the heavenly kingdom. The baptism cleansed the initiate and allowed him or her to share in the death and resurrection of Christ. The Mass performed a similar function. The offering of the eucharistic sacrifice allowed the individual to take part in the Last Supper and the subsequent death of Jesus.[15]

Paul preached to the faithful that the new order had already arrived with the life and death of Christ. While the historical Jesus might have come and gone, his spirit was now always present, ready to redeem fallen souls. Paul held up a new image of the future in which the Jewish expectation of salvation someday was replaced with the offer of salvation now. Instead of an earthly deliverance, Paul preached deliverance into a new spiritual kingdom. The Christian image of the future allowed the individual to experience "the Kingdom" in the here and now. By accepting Christ as their Savior, the faithful would still be in the world, but not *of* the world.[16]

Later on, St. Augustine concretized this new vision with his division of the world into two spheres, the earthly city and the city of God. The true Christian lived in both spheres simultaneously. While he was expected to render to Caesar that which is due, the Christian's spiritual allegiance was always to be directed to Christ the Redeemer. St. Augustine made it very clear that he looked upon the earthly city as a temporary evil, an irksome reality only to be tolerated on the road to ultimate salvation in the kingdom of heaven after death. In the meantime, every Christian could participate, at least in part, in that expectant perfect order by being born again into the heavenly kingdom on earth, Augustine's City of God.[17]

To enter into the City of God, the initiate must do more than just be born again in Christ. It was not enough to be baptized and to take part in the Mass. One must also surrender fully to Christ's teachings. In this respect, the new image of the future, as expressed in the teachings of Christ and the Church fathers, was unlike any other prescription in Western experience. To take part in the new spiritualized kingdom, the faithful must turn away from the old way of life and adopt a new approach to their fellow human beings that called for complete self-sacrifice. To turn the other cheek, to love one's enemies, to never take up the sword, to minister to the poor, are all essential guidelines for living a Christian way of life. To live in Jesus, then, is to live as he did.[18]

This otherworldly image of the future dominated the politics of the Holy Roman Empire during the long medieval era, providing a unifying vision for all of Western culture. Frederick Polak, writing of this "New Image of the Man in Jesus," observes that in the emergence of this body of teachings, Western civilization is presented with a new spectre, a new vision of the future, and a new order of perfection toward which to strive.[19]

The Church fathers, the priests, bishops, cardinals, and pontiff became the new intermediaries with the future. They held in sacred trust the image of the future that Christ, and later Paul, extended to the world. The virtual monopoly they had over Latin script assured that they alone would be in a position to interpret God's will as expressed in the Bible. The priests of the Church also controlled the baptismal ritual and the Mass. Control over these functions assured that they alone would have the power to grant access to the new image of the future that Paul extolled.

While every man and woman could choose to accept Christ as their Savior, the Church fathers determined who would be al-

lowed to be reborn in Christ because they controlled the transforming rituals. The priests of the Church stood squarely at the
door that opened up into the new Christian image of the future
and, from this vantage point, exercised complete hegemony over
the future vision of Christianity.

The Church claimed to be God's designated outpost in a fallen
world. It granted to itself ultimate authority to determine God's
will and to address the means by which the faithful might enter
the kingdom. It also took on the mission of preaching the gospel
to the infidels, and of spreading the good news that Christ died
to cleanse humanity of its sins. The Church marshaled its resources in the service of saving fallen souls for Christ. According
to Church doctrine, the Second Coming of Christ would follow
in the wake of the Church's successful campaign to convert the
nonbelievers. Therefore, even the image of the future that Paul
had proclaimed depended to some degree on the successful intervention of the Church if it was to be realized.

History is replete with examples of competing individuals and
groups attempting to impose their own unique image of the future
onto the body politic. Hebrew prophets and priests, as well as
Christian clerics, have intuitively grasped the age-old lesson of
political power. Power always flows to those who grab possession
over the society's future. They are the select few who have,
throughout Western history, claimed the right, authority, and
wisdom to be able to predict, forecast, and intervene with the
future on behalf of the people.

Images of the future are the single most powerful socializing
agents that exist in Western culture. We know this to be the case
because whenever a strongly adhered-to image of the future has
been pitted against sheer force of arms, it has eventually triumphed.
It is true that states have been able to exist for brief moments by
use of brute power alone, but, without an image of the future to
sustain them, they quickly become consumed in and exhausted
by their own violence. The great states and civilizations in history,
those that have endured the test of time, are precisely those that

have possessed a compelling image of the future, an image strong enough to seduce, capture, and sustain the energy and commitment of succeeding generations.

It has been said that "the rise and fall of images of the future precede or accompany the rise and fall of cultures."[20] The great political battles in Western history have been waged over competing temporal visions. The human family has locked horns over questions of temporal orientation through most of recorded Western experience. If this proposition is difficult to accept, perhaps it is because the age we live in is so steeped in a materialist consciousness that we cannot possibly imagine the idea that time considerations might play as important a role in the political process as economic considerations.

Nowhere is the materialist bias more evident than in the writings of Karl Marx. Like many of his contemporaries, Marx viewed material reality as the only reality. He argued that changes in material conditions give rise to new modes of production that, in turn, change the way people think and act in relation to each other in a social setting. Marx made the mind a hostage of the body and the body a hostage of the material conditions that provide it with sustenance. Changes in consciousness, Marx argued, follow changes in both the material conditions and the modes of production. According to Marxist eschatology, history unfolds in a deterministic manner, leaving little room for the human mind to make its own choices regarding alternative futures. Material conditions, Marx argued, inevitably dictate changes in economic and social thinking and changes in political relationships.

The mind, however, is more than a passive captive of its physical and material surroundings. The mind is continually initiating change by fashioning new temporal orientations, and these new time dimensions interact with the changing material conditions to help shape the context for the kind of economic, social, and political realities that emerge. Temporal transformations accompany and, on occasion, even precede material transformations. Without changes in the consciousness of time, the great economic and political changes in Western history could never have taken place.

The most recent past example of the workings of temporal transformation is found in the transition from the medieval to the industrial era. During this period, Western civilization underwent a complete metamorphosis in time consciousness, paving the way for the eventual adaptation to and acceptance of a new mode of production and a new and powerful image of the future.

10

THE IMAGE
OF PROGRESS

The struggle between the medieval Christian church and the emerging bourgeois class of merchants and artisans was, to a great extent, a struggle over competing temporal orientations and, ultimately, a struggle over different images of the future. As far as the Church was concerned, earthly time was not of much relevance. If anything, time was conceived of as a necessary evil, something that had to be endured. For the devout Christian, this worldly existence was meant to be spent in preparation for the eternal life that awaited after death. There was never a question of using time to improve one's lot or the well-being of the society. Indeed, such thoughts might have been regarded as heresy. To entertain the notion that this world could be improved on was, in the mind of the Church, a sin of pride. After all, God in his infinite wisdom had designed his earthly creation as he would have it. Any man or woman who dared to challenge God's artistry by attempting to effect changes risked the wrath of the Church and the Divine Power.

In medieval Europe, the Church had established a proper order for the functioning of every aspect of social life, leaving little room for change. Everything had its appropriate place and role in the Christian scheme of things and, as Frederick Polak points out, "to rebel against the place assigned by God to Man in this world would be to commit a deadly sin and challenge the very throne of God."[1]

134

While the Christian believer was firmly ensconced in the world of the flesh, his or her heart, mind, and soul were always fixed on the heavens above, where salvation awaited. Because this image of the future rested so little on the mundane fortunes of earthly existence, the passage of time was never a matter of great concern or overwhelming import. The Church formalized this otherworldly image of the future by praising all heavenly pursuits and denigrating any purely secular endeavors, especially those that threatened to alter the economic, social, or cultural landscape.

The Church published a list of forbidden and dishonorable trades. "Virtually all Medieval professions" were, to one degree or another, deemed unfit, unclean, and unacceptable as they were all associated with "the ways of the flesh." Only agriculture and a few select trades—including goldsmiths, ironsmiths, and swordmakers and, of course, the clergy itself—were spared the condemnation of the Church hierarchy. While bathkeepers and innkeepers were condemned for promoting licentiousness; butchers, fullers, dyers, and cooks were chastised for being unclean; and surgeons and barbers were eschewed for spilling blood; the Church reserved its greatest contempt for the merchant class.[2]

Man's work was supposed to be in the image of God, and, as God's work is creation, any profession that did not create something tangible was to be condemned. The Church attacked the rising merchant class as a parasitic force, a group of conspirators who created nothing of value and only exploited the work of others.[3]

Yet the Church's condemnation was not enough to stem the tide of a burgeoning commercial revolution that was fast extending its reach over the length and breadth of the Holy Roman Empire. By the thirteenth century, the merchant class was beginning to tilt with the Church in a struggle to wrest control over the secular affairs of the fast-growing urban sectors of the continent. While power was the central issue, it was the differing concepts of time that locked the two contending forces into a bitter ideological confrontation.

For the merchants, time was everything. Their success or failure depended upon their ability to use time to their advantage. Knowing when the best time was to buy cheap and sell dear; how long inventory should be allowed to stay on hand; determining the time it would take for goods to arrive, or how long it would take to ship them to their destination; being able to time changes in exchange rates, the rise and fall of prices, changes in labor availability, the time necessary to make a product, were all critical factors. The merchant who garnered the most knowledge of how to predict, use, and manipulate these various time frames commanded the best prices and made the most profit.[4]

The merchants' use of time had played virtually no role in the affairs of Europe during the long medieval era. Cut off from any significant commercial trade for nearly eight centuries, economic life centered around the self-sufficient and self-contained manor estates. Whatever trade existed between communities was extremely limited and barter, rather than monetary exchange, was the rule of hand. In the medieval economy there was no need to think of time as a scarce resource that could be manipulated for personal economic gain.

For the Church, time marked the interval between the first and Second Coming of Christ. Time was viewed as a waiting period, something God freely granted in order that the faithful might prepare for the coming of the Lord and eternal salvation. Time was something God disposed of—it never would have dawned on the prelates of the Church that human beings had the right to use or exploit time as if it were something they could rightfully possess or control.

The Church's concept of time was, of course, at direct odds with the merchants'. While the merchants argued that "time is money," the Church contended that "time is a gift of God and therefore cannot be sold."[5] The Church developed an elaborate code of ethics to insure against time profiteering by the merchant class. For example, consider the question of whether a merchant was entitled to "demand a greater payment from one who cannot settle his account immediately than from one who can?" According to the Church, "the answer is no, because in doing so he would be selling time and would be committing usury by selling what does not belong to him."[6] The Bible was clear on the matter. In Luke 6:35, Christ says, "Lend hoping for nothing again."[7] Because they profited by using time to their advantage, the merchants were looked on as sinners. "Their profit implied a mortgage on time, which was supposed to belong to God alone."[8] By steadfastly refusing to accept the new notion of time as an exploitable medium, the Church set the terms for an inevitable showdown with the increasingly powerful merchant class.

The increase in population, the emergence of new urban centers, the opening of transportation routes and communication channels with the outside world for the first time since the days of Roman rule all led to a confrontation over two different economic concepts and two different concepts of time. Jacques Le Goff sums up the impact of the struggle:

> The conflict, then, between the Church's time and the merchant's time takes its place as one of the major events in the mental history of these centuries.[9]

For the Church, time marked the passage from this world to the next. For the merchants, time was a tool to advance the interests of Mammon. The Church gave little notice to earthly time while the merchants reduced time to money. In the end, the merchants' concept of time would win out, but not without a long and protracted struggle with the Vatican.

The Church's concept of time went down to defeat largely because its image of the future was not strong enough to withstand the tremendous changes that were reshaping the tenor and tone of European life. Between the thirteenth and seventeenth centuries, Western Europe was ravaged by disease, racked by economic dislocations, and torn asunder by a wave of political insurrections. The fall of the medieval image of the future and

the meteoric rise of the modern worldview can only be understood
in the context of the tumultuous events that turned Europe up-
side down over the course of four centuries. The ensuing confusion and malaise served to undermine confidence in the prevailing image of the future and opened the door to a new future
vision that would eventually succeed as the ruling paradigm in
the European theater.

Many factors contributed to the transformation of European
culture between the thirteenth and seventeenth centuries, but
none had more measurable impacts than the Black Plague,
the crises in agriculture, the enclosure movement, and the
population migration from rural areas to towns and cities. Each
of these events unsettled traditional patterns of medieval life,
creating a host of new tensions that the Church worldview was
ill-prepared to address. Questions mounted faster than the clergy
could provide convincing answers, causing a loss of confidence
in the Church's image of the future.

Beleaguered by mounting self-doubt over an old worldview that
no longer seemed to work and still without a new image of the
future to take its place, Europe sank deeper and deeper into
despair. The future appeared bleak and dim; life became stripped
of purpose and meaning. Commenting on the time in which he
lived, Pascal wrote:

> Ever drifting in uncertainty, driven from end to end, man feels
> his nothingness, his forlornness, his insufficiency, his de-
> pendency, his weakness and there will immediately arise from
> the depth of his heart weariness, gloom, sadness, fretfulness,
> vexation, despair. . . .[10]

No longer convinced of his prospects for security in this world
or immortality in the next, European man was becoming increasingly consumed by fear over what lay ahead. The schedule and
the clock became a means to divert attention from the vagaries
of the future. William Bouwsma speculates that the proper regulation and use of time eliminated some of the uncertainty of life.

Scheduling time became a way to ward off anxiety. By planning every detail of life in advance, it was possible to fill in the future in such a way as to leave no time for uncertainty to intervene. Slowly, the bourgeoisie began to advance the idea of securing the future through the proper husbanding of time. The clock became the instrument to hoard and mete out time. The clock's introduction into the economic life of Europe led to the idea that time could even be bought and sold. The traditional notion of selling one's labor or skills was replaced with the new concept of selling one's time. The hourly wage and the piece rate helped to establish the idea that time is money.

If time could be bought and sold in units, it could also be accumulated or depleted. In the new clock culture, time and money became interchangeable and exchangeable. The more money one could amass, the more time one could buy and sell. Whereas medieval man and woman had believed that accumulating "good works" would help assure them security in this world and eternity in the next, the new bourgeois class was expressing a quite different belief—that the accumulation of time and money was the best means of providing both security and a new form of earthly salvation. In succeeding centuries, the bourgeoisie would comfort themselves in the belief that they had it in their power to extend duration indefinitely in this world by the proper management of time.

It is within this context that a powerful and compelling conception of the future emerged, laying the philosophical foundation for the modern worldview. The great intellectual thinkers of the day began to espouse the radical idea of human progress, a wholly new vision of the future for Western civilization to rally around.

Francis Bacon opened up the floodgates to this new image of the future in 1620 with his *Novum Organum*. He called for a new approach to organizing the world, one that could "enlarge the bounds of human empire, to the affecting of all things possible."[11] He called his new approach the "scientific method," and said that its proper use would allow people to take "command over natural things—over bodies, medicine, mechanical powers and infinite others of this kind."[12]

Confident that the universe was, indeed, an orderly proposition, Bacon came to believe that he could uncover its operating principles in much the same way a master clockmaker might be able to take apart a clock and analyze the parts in order to learn how it operates.

Bacon was convinced that the clocklike order of the universe could be discovered by what he referred to as "objective" thinking. The goal, according to Bacon, is to be able to predict how nature will act and unfold so that the future can be brought under control. Bacon's approach to the future was quite different from St. Augustine's. Bacon was more interested in controlling the immediate future in this world. St. Augustine was more concerned with gaining access to the "other" world. Bacon believed that the scientific method offered the best vehicle to predict and control the temporal horizon. St. Augustine argued that only God's emissaries—the priests of the Church—were equipped to predict what lay ahead. Bacon was interested in material security and looked to the prospect of advancing steadily toward a cornucopic kingdom here on earth. St. Augustine was interested in spiritual salvation and looked to the Second Coming of Christ and eternal salvation in heaven. For Bacon, the future depended on the exercise of human will. For St. Augustine the future depended on God's grace.

Francis Bacon confronted the Church's vision with a heretical notion, proclaiming to the world that human reason, not divine wisdom, would fill in the outlines on the temporal horizon. The God-centered image of the future, which had dominated the pages of Western history, was being challenged by a new man-centered image. From now on, the old motto, Pray for spiritual guidance, was replaced with a new motto, *sapere aude,* or, Dare to think.

Armed with the scientific method, the architects of the Enlightenment headed toward the temporal horizon, determined to

lead the way to an earthly paradise. René Descartes helped blaze
the path to Eden by introducing mathematics into the new for-
mula for the future. Descartes envisioned the giant cosmic clock
as a precise, orderly, mathematical model and proclaimed that,
by using mathematics as our primary form of knowledge, it would
be possible to elicit "true results in every subject."[13] Descartes
was convinced that mathematics could unlock the secrets of nature and thus make human beings the rulers of the universe.
"To speak freely," he said, "I am convinced that it [mathematics]
is a more powerful instrument of knowledge than any other that
has been bequeathed to us by human agency, as being the source
of all things."[14]

The revolution in temporal thinking was a long time in coming.
Paul had offered faith as the key to paradise. Twelve centuries
later, St. Thomas Aquinas had appended Paul's enthusiasm by
arguing that faith and reason together were the stepping-stones
to the ideal future that awaited humankind at the end of its
journey. Now Descartes dismissed faith altogether and called
upon future generations to rely on quantifiable reason alone,
unadulterated by the superstitions of the past.

Bacon and Descartes believed that "knowledge is power." By
relying on the scientific method and rigorous mathematical logic,
they argued that it was possible to both understand and control
the forces of nature, and, by so doing, advance the material well-
being of society and create a more secure world.

The Hebraic and Christian cosmologies introduced history into
their images of the future. Bacon, Descartes, and their contemporaries introduced progress. This was a revolutionary new idea
for which there was little precedent. Time, in the new scheme
of things, was no longer to be used in preparation for the Second
Coming of Christ but rather as a means to advance the new
temporal idea of progress. To believe in progress is to believe in
a future that is always improving, enlarging, and, above all, en-
during. There is no end to progress. It is unstoppable, relentless.
It speeds us into a future where there are no boundaries or bor-

ders, a future that is infinitely expansive and "timeless."

This new image of the future informs the modern mind. It is an image steeped in materialism. Material progress is our ticket to immortality, our way of cheating death, of overcoming a fleeting existence. The more material we can amass, the more confident we become that the cornucopic vision we have laid out before us is reality, not illusion, a real place, not an imaginary figment.

The philosophers of science provided the tools for amassing a materialist cornucopia, but it was the political philosophers and the economic theorists who provided the rationale and the will to ensure its proper welcome. John Locke argued that material self-interest is the underlying motivation of all human behavior. The proper role of a government was to ensure that the forces of nature were brought under control and harnessed so that each member of society could accumulate as much wealth as humanly possible. "The negation of Nature," Locke contended, "is the way toward happiness."[15] Adam Smith concurred, and offering a bit of economic wisdom to bolster the argument for a materialist vision of the future, said:

> Every individual is continually exerting himself to find out the most advantageous employment for whatever capital he can command. It is his own advantage, indeed, and not that of society which he has in view. But the study of his own advantage naturally, or rather, necessarily, leads him to prefer that employment which is most advantageous to society.[16]

In the works of John Locke, Adam Smith, and their contemporaries, we see the unfolding of a new model of the future, one that redefined the time horizon for successive generations of Europeans, eventually becoming the dominant future vision of much of the world community. Henceforth, "material progress" would be the raison d'être of the age.

Not surprisingly, political and economic leaders looked to the scientists and mathematicians for help in forecasting and controlling this new progressive vision. In place of astrological charts and crystal balls, revelation and divine mediation, the scientific elite placed before the world the combined power of the scientific method and mathematical reasoning. The power brokers of the modern age have come to rely almost exclusively on the tools of science and technology to secure the future.

Historian Stephen Kern acknowledges the debt we owe to the scientific method and mathematical measurement:

> Gravitational astronomy is able to predict stellar movements, medical science continually improves its ability to diagnose, meteorology predicts the weather and chemists forecast elements before they are discovered, as Clerk Maxwell announced the existence of rays before Marconi put them to use.[17]

Science and technology have become the new means of obtaining salvation. In times of crisis we look to science and to the technological products of science to rescue a fallen humanity from the foibles and follies that so often intrude on our efforts to eke out a safe domain. Modern science and technology are the secular messiahs in a materialist world. They are the guarantors of our security and, ultimately, our immortality. Through science and technology we will extend our control over the future, the forces of nature, and our own bodily duration. We will live better, live longer, enjoy the good life, and enter into an earthly Eden of our own making where material abundance will provide a fortress against the ravages of time and the onslaught of death.

The new bourgeois image of the future seemed invincible. Human beings now believed that the more material possessions they could amass, the more time they could buy for themselves. Everywhere they looked there were signs to confirm their new progressive vision. The whole world seemed to be improving—and expanding.

European explorers had opened up the world for colonial rule. Ships were shuttling back and forth across the seas, extracting resources from the four corners of the earth and subduing entire populations and continents in the name of progress. (Regular oceanic travel would not have been possible without the watch; it provided the ship's captain with a means of judging longitude on the high seas.)

With the world in their hands, European merchants, traders, and political leaders rushed to convert the planet's vast resources into products that could be sold in the ever-expanding markets being brought under their control. Steam engines were hustled into place all across Europe, providing a new source of energy that promised to replace animal power, manpower, and even the vast energy of the sun. The steam engine "was outside the varying seasons, did not recognize the difference between night and day [and] imposed its own rhythm on the men who worked for it."[18] The steam engine greatly accelerated the conversion of raw resources into finished goods. Overnight Europe and the world were awash in a sea of new products. The material cornucopia seemed just around the corner.

The transition from agriculture to industry was accompanied by a second wave of migration from country to city. People streamed into the new urban industrial centers in overwhelming numbers. In 1801, more than 80 percent of the population of Britain still lived in the countryside. Just fifty years later, only half remained.[19] The rest had picked up their belongings and resettled in the bustling new industrial cities of Manchester, Leeds, and Liverpool.

The train made its debut in 1825, chugging along a track stretching from Brussel to Stockton, England, at a blistering thirty miles per hour, "trebling the best land speed yet regularly attained."[20] Sixty-five years later, the electrified underground made its first run in London, and regular passenger rail to and from the city became standard fare, giving rise to the "suburb" and rush-hour traffic congestion.

The suburbanite epitomized the clock culture.

In every dormitory suburb, over the length of each road, within the same brief period in the morning, a hundred doors would open, and a hundred breadwinners emerged, like automata from a medieval town clock, to converge on a station where a hundred watches would confirm and a hundred voices agree whether the train was, or was not, on time.[21]

The train schedule became the regulator of daily life. To insure punctuality, on-time performance, and strict adherence to the train schedule, all rail personnel were equipped with special train watches that "would not gain or lose more than 40 seconds in two weeks and had to be cleaned and regulated twice a year by a railroad watch inspector."[22]

After centuries of near isolation, small rural hamlets were being connected with each other and the world through the establishment of national postal services in each country. For centuries postal delivery had been sporadic, unorganized, and unpredictable. The clock and the schedule changed all that in dramatic fashion. By 1863, London could boast of eleven mail deliveries per day. A letter "posted early in the morning to another London address could not only bring a reply, but do so in time for a further letter from the first writer to be delivered before the day was out."[23]

The teletype, telephone, and daily newspapers broadened the span of communication and narrowed the distances separating people. Events across the country and around the world could now be overheard and be made an object of gossip and speculation from the backyards of bricklayers to the courtyards of kings. The speedup in communications was accompanied by a compulsion to know what was happening everywhere at all times. Thoreau wrote, "Hardly a man takes a half hour nap after dinner, but when he wakes he holds up his head and asks 'What's the news?' "[24]

The telegraph and the telephone also accelerated the pace of economic activity. Commercial decisions could be made virtually instantaneously. Markets could be coordinated over vast geographic spaces simultaneously.

These new "time-saving" inventions even changed the complexion of national and international politics. Diplomacy, once a slow and ponderous activity, was now subject to "crisis" resolution as the new forms of communication shortened and intensified the time frame for political debate and resolution.[25]

In 1809, London extended day into night with the introduction of the first gaslight. The new invention lit its way across the continent, turning the shadowy underside of city life into a dazzling array of color that seemed to mirror the twinkling of the northern lights.[26] In 1879, Thomas Alva Edison invented the electric light, further hastening the process of lighting up the night skies. The famous historian of architecture, Rayner Banham, heralded the new invention as "the greatest environmental revolution in human history since the domestication of fire."[27] Night was now interchangeable with day, and humanity was convinced that with this extension they had bought themselves great quantities of extra time.

In a span of just two hundred years, a revolution in time had taken place. Durations were dramatically shortened; the sequencing of activity sped up to near simultaneity; activities were timed to the hour, the minute, and even the second. Events spread over vast geographic spaces were synchronized by clock and schedule to exacting standards. The clock culture and the schedule issued forth a new dawn. Accelerated energy, accelerated speed, accelerated communication, accelerated material accumulation—could anyone doubt that progress was on the march, that we could indeed reach the earthly cornucopia of Bacon, Descartes, Locke, and Smith?

The new future lay before the human family, a wonder to behold, a prize to be secured. It would be ours for the taking. No impediment, no obstacle would be allowed to stand in the way of progress. Here was a new and unflappable force, resolute and invincible. It would sweep us and our children into the promised land. Our course was set, our destination assured.

Western civilization was thrust into a new time orientation, complete with a new set of symbols and images. The new god was science and technology; the new salvation, material progress; the new church, the industrial order; the new idol, the clock and watch; and the new ritual, the daily schedule.

At the beginning of the modern age, when the exuberance of this newfound future image was first impressing itself on European consciousness, the French aristocrat, the Marquis de Condorcet, captured the sense of euphoria that was sweeping over the intellectual community in words that have since become immortalized in history:

> No bounds have been fixed to the improvement of the human faculties . . . the perfectability of Man is absolutely indefinite . . . the progress of this perfectability, henceforth above the control of every power that would impede it, has no other limit than the duration of the globe upon which Nature has placed us.[28]

11

THE VISION OF SIMULATED WORLDS

The image of progress dominated the time politics of Western culture for nearly two centuries. It has provided the motivational context for mobilizing the temporal affairs of individuals, communities, and nations. Now as computime begins to accompany clock time and programs begin to subsume schedules, a new temporal orientation is beginning to emerge, and along with it a new image of the future. The age of progress is about to give way to the age of simulation. The new simulated vision of the future incorporates the Promethean appetites of the former age while rejecting the restraints that have shackled it to historical consciousness. History barely exists in the new image of the future. The new vision is influenced more by popular psychology. The future is no longer viewed as something that unfolds in strict linear fashion along a historical plane. Rather, it is something that is continually being reprogrammed to suit the transitory needs of each emergent reality. Historical terms like "fate" and "inevitability," which so dominated the thinking of the age of progress, are being replaced with psychological terms like "choices" and "scenarios" as we make the transition. The new future image conceives of reality as a vast reservoir of information to be fashioned into simulated experiences.

In the Bible it is written, "In the beginning was the Word."

148

God created the world *ex nihilo*. He thought it into existence, then he spoke it into reality. He said, Let there be light, and so it came to be. The new image of the future also begins with the word. The word is represented in coded messages, pieces of information that can be edited together to create thoughts, ideas, and activities. With information, we can turn chaos into order, darkness into light. We can design new worlds of our own making. These new worlds flow directly from the psyche. They are simulations born of pure thought. They are the Edens of our own imagination. We now can free ourselves once and for all from the restraints of a former creation. We now have it in our power to reduce matter to energy and energy to information. We can tear away the boundaries that have given form and substance to reality and reimagine the world made up solely of information, of messages, and of instructions. Such a world is boundaryless; it can be rearranged at will, programmed to flow in and out of forms, edited into countless new structures.

Man and woman, made in God's image, now set out to make new worlds in their own image. These are simulated worlds, but nonetheless impressive for being so. The fact that they are erected with the aid of an artificial intelligence is of little concern, for the new man and woman have come to believe that "a perfect simulation of intelligence is intelligence."[1] Human beings now dream of unbounded creations stretching into a future without end. Man, the agent of change, is about to become man, the creator of worlds.

Alvin Toffler speaks to the new image of the future when he says that humankind will increasingly "own the technology of consciousness."[2] With the computer, humanity creates a simulation of its own psyche, a second mind force that can be let loose on the world to rethink, reimagine, and redirect its course.

In Talmudic lore, God is said to have made several different attempts at fashioning the world before settling on the one we have come to inherit. The new image of the future likewise envisions many worlds, many creations.

The programmer-God makes the world not once and for all but many times over again, rearranging its elements to suit each new program of creation. The universe proceeds like a program until it runs down or runs wild, and then the slate is wiped clean, and a new game begins.[3]

In the age of simulation, the vision of a one-dimensional linear history is replaced with the image of programming limitless future realities. In the new scheme of things, there are as many forms of paradise as there are new realities to program.

The purveyors of this new image of the future are anxious to move society onto the center stage of a new world drama where we become writer and director as well as actor. For those who might entertain second thoughts about the new course being chartered, the proponents of the age of simulation are quick to issue a word of warning. Speaking for the committed, Edward Feigenbaum admonishes the doubters, the faint of heart, all those who hesitate to make this journey:

Those intellectuals who persist in their indifference, not to say snobbery, will find themselves stranded in a quaint museum of the intellect, forced to live petulantly, and rather irrelevantly, on the charity of those who understand the real dimensions of the revolution and can deal with the new world it will bring about.[4]

That new world was spawned in the intellectual cauldrons of Western thought, but is being transformed from thought to reality far to the east in places like Taiwan, Singapore, and Korea.

The center of the new computer time world is Japan, which has made a national commitment to be the first fully computerized information economy by the early decades of the twenty-first century. The government, working in tandem with private industry, has poured funds into developing a new fifth generation of computers—machines that will far exceed earlier models in their ability to "simulate" human thought. These will differ substantially from the electronic vacuum tube computer, the transistorized computer, the integrated circuit computer, and the emerging large-scale integrated computer. The first four generations of computers collected, stored, and processed increasing flows of information in shorter time spans. The new fifth-generation computer will be far speedier. What is more important, say researchers, is that it will also have the capacity to reason, not just compute.

Proponents of fifth-generation computers are eager to make the distinction between earlier computers and the new "reasoning" machines, and have gone so far as to call the new machines by a new name. In the trade, fifth-generation computers are referred to as KIPS, Knowledge Information Processing Systems, in an effort to distinguish them from the idea of a computer as a mere computational or counting machine. These machines "signal the shift from mere data processing, which is the way present-day computers function, to an 'intelligent' processing of knowledge." The new machines, according to the engineers who are designing them, will be capable of exhibiting artificial intelligence.[5]

Many industry analysts expect the fifth generation of computers to push much of the world community inextricably into the information age sometime early in the next century. Futurists talk in glowing superlatives about the new world that waits in the offing. The new economic order, they contend, will build on and eventually pass over the older industrial way of life.

> The wealth of nations, which depended upon land, labor, and capital during its agricultural and industrial phases . . . will come in the future to depend upon information, knowledge, and intelligence.[6]

While the proponents acknowledge that traditional economic needs will still have to be met, they argue that harnessing energy, making products, and feeding people will be subsumed by a larger dynamic in which economic planning and implementation will be a function of the new information order.

> In control of all of these [economic] processes will reside a new form of power which consists of facts, skills, codified experience, large amounts of easily obtained data, all accessible in fast, powerful ways to anybody who wants it—scholar, manager, policymaker, professional, or ordinary citizen.[7]

In the new world, "knowledge itself is to become the new wealth of nations," a "saleable commodity like food and oil."[8]

The new economic vision is only part of a larger vision that is beginning to form around the computer time world. The transition from the age of progress to the age of simulation is not only effecting a change in the way people see their economic futures, but also their psychic and spiritual futures as well. In 1972, the Japan Computer Usage Development Institute presented the government with a report entitled "The Plan for an Information Society: A National Goal Toward the Year 2000." The plan, which has since been adopted by the Japanese nation, is authored by Yoneji Masuda, a leading figure in the computer field. Masuda spells out, in some detail, the new image of the future he envisions as a companion to the emerging computer revolution.[9] Of particular interest is the fact that Masuda identifies time as the first and most important value of what he refers to as the coming "computopia."

"My first vision of computopia," says Masuda, "is that it will be a society in which each individual pursues and realizes time-value."[10] Masuda defines "time-value" as "painting one's design on the invisible canvas of one's future and then setting out to 'create' it."[11] Masuda sees time-value as subsuming the material values of the older clock culture. While material benefits will still be maximized in the new computopia, they are considered to be of lesser importance in the ultimate scheme of human evolution. According to Masuda:

> Time-value is on a higher plane in human life than material
> values as the basic value of economic activity. This is because
> time-value corresponds to the satisfaction of human and in-
> tellectual wants, whereas material value corresponds to the
> satisfaction of physiological and material wants.[12]

Time takes on a new importance in the computer time world,
says Masuda, because of "the increased effectiveness of purposeful action."[13] With computer technology, it is possible for the
first time to foresee various futures and devise logical programs
in advance to make those futures come true. The new computer
timepiece allows us to gain greater control over the futures we
envision.

During the age of progress, material production was the prized
value. In the emergent simulation age, mental production becomes the *sine qua non* of advancement. Social change is no
longer measured by the organization of matter and energy into
larger, more self-contained structures but in the processing of
data into more complex interrelated systems of information.

The proponents of the new image of the future equate information enrichment with mental improvement. In the new computer time world, evolution of information and evolution of
consciousness become interchangeable and tautological. The
universe is no longer pictured as a giant cosmic clock ticking
away in predictable, orderly fashion in compliance with the Newtonian laws of mechanics. The new universe resembles a giant
computerlike mind, ever expanding, creating new information
and new knowledge, filling the cosmos with higher and higher
levels of consciousness.

Many computer programmers identify with the new view of
an evolving universal consciousness. They see their own work
in artificial intelligence as a creative act, adding increments of
information to the expanding mind force of the universe. Sociologist Sherry Turkle says of this new priestly class that they
"relate to one another not just as technical experts, but as creative

artists."[14] They are engineers of the mind, creators of new forms of simulated intelligence. Their mission is quite unlike the one that motivated the best and the brightest in the clockwork culture, who toiled in search of more efficient means of increasing material output. Their successors gaze toward a far different horizon. "They lose themselves in the idea of mind building mind and in the sense of merging their minds with a universal system."[15] The cornucopic vision is being subsumed by the computopian vision.

In the coming new world, both the role of religion and the computer engineer's concept of God are changing fast, and like almost everything else, the models are inspired by cybernetics. God's role is becoming more contemporary. The new God is no longer the Creator but merely the mind of the universe—an informational diety that can be tapped into with increasingly more sophisticated computer programming techniques.

Human beings reach out to this new informational deity by interfacing with the evolving mind of the universe. Communion is the experience of gaining access to larger and larger stores of information, of simulating more complex programs, each reaching ever closer to the ultimate computerbank storage facility, the mind force of the cosmos.

Like the age of progress, the new simulated image of the future is open-ended. However, while the former age is constrained by history, the new age is not. In the age of simulation, the past is made ahistorical. Experiences are severed from their historical context and transformed into tiny bits of data that are suspended in a timeless setting. In this sense, part of the new computer world is like a simulated version of Freud's unconscious. Freud defined the unconscious as a timeless realm of the mind, a place full of inchoate experiences and unconnected thoughts, free-floating in a kind of chaotic collage. The computer analogue to the unconscious is raw data, pieces of experience disconnected from their historical context, made into bits of information, waiting to be reordered and put to use. The central processor of the computer is a facsimile of consciousness. It programs the unconnected raw data into a set of purposeful instructions or actions. It removes data from a timeless setting and imprints temporality into it. The computer past is not fixed, linear, or chronological as in the old clock culture. The new past is formless, timeless data, taking on new meaning every time it is transferred into new programs. The new time is associative rather than linear. It is a stepchild of psychological consciousness, just as the concept of linear time was a stepchild of historical consciousness.

Everything in the new computer world is temporary and fleeting. Everything is subject to continual edits, revisions, and modifications. Time loses the independent status it enjoyed during the clock culture. Time is now a resource, not a reference point. It is information to be shaped in countless new ways, imprinted with new meanings along the way. David Bolter captures the sense of impermanency that pervades the new computer culture:

> A programmer can never forget that every solution in the computer world is temporary, makeshift, obsolescent. . . . The ease of transferring data and the huge amounts of data transformed mean that nothing in the computer world remains long in one form.[16]

There is no well-established past, no preconceived future, no starting point or end of the line in this new world, just the unceasing process of simulation. Even the new frontiers being programmed are fleeting, each simulation enjoying a moment of existence, only to be quickly edited off the screen of consciousness to make room for still another program. The new image of the future is more cathartic than cumulative. Bolter grasps the significance of this monumental change in temporal awareness, from historical to psychological consciousness, when he notes that the computer man "does not speak of 'destiny' but rather of 'options.' "[17]

In the fast-moving time world of the computer age, realities change at such accelerated speeds that even scientific truths become impediments and ultimately become expendable. In all other periods of history, knowledge came slowly. For that reason, each new insight was enshrined and closely guarded. It was elevated and made timeless. It became something to live by. It lasted because great lapses of time occurred between insights. That is no longer the case. What we know today is quickly eclipsed by what we know tomorrow. Thus we can no longer tolerate timeless truths and ironclad laws. By their very nature, timeless truths and ironclad laws impose boundaries, and it is for this reason that they are now dispensable. Timeless truths and ironclad laws tell us what is not possible. They establish upward limits to what we can do. They serve as a notice as to how far it is possible to go.

With everything changing so fast it is necessary to construct a temporal cosmology in which change is honored as the only timeless truth. By reinterpreting nature as the evolution of information, humanity achieves this end. Nature is no longer seen as a set of restraints but rather as a process of creative advance. In this new scheme of things, even the laws of science lose their potency. They are no longer seen as truths but merely as convenient instruments to advance the information process. Nothing is considered permanent except the ongoing process of information gathering and processing.

The new image of the future also brings with it an ethic that is more congenial to the sense of impermanency that pervades the new computer time world. In the medieval and early industrial eras, ethics was conceived of as a set of absolute principles that existed a priori to human experience. In the past two centuries, the timeless nature of ethical precepts has gradually eroded with the ascension of utilitarianism and, more lately, situational ethics. As we enter into the computer age, ethics is about to transcend its timeless setting altogether, becoming as mobile and facile as the accelerated time frame within which it will operate. Philosopher and cosmologist Eric Jantsch says that ethical behavior in the coming information age will be "behavior which enhances evolution."[18] The new evolutionary ethics, says Jantsch, will "explicitly include the main principles of evolution such as openness, non-equilibrium, the positive role of fluctuations, engagement, and non-attachment."[19] Change for change's sake becomes the new ethic. Jantsch argues that humanity's highest responsibility is to spur on the evolutionary process: "Creative processes ought to be permitted to interact freely and to find their own order of evolving structures."[20]

Up until the modern era, every culture in history placed a premium on customs and traditions. The clock culture and the computer time culture shift the ethical canons from past to future and from restraints to resourcefulness. Before the modern era all ethics used to begin with "Thou shalt not." During the industrial age, the ethical code shifted to the idea of "Be productive." The ethics of the computer time world is "Be creative." Evil in the information age is viewed as anything that places obstacles in the way of the innovative act, the novel experiment, the untried scenario.

Jantsch envisions a new type of manager-priest to oversee the coming age. More inquisitive than inquisitor, more of an innovator than a disciplinarian, the new leader in the emerging computime world would be a catalyst for change.

> His task would be primarily the prolongation of those processes which seem to run in a creative direction, to stop those which appear unpromising and eliminate those which he deems uncreative. At the same time, it would be his task to stimulate and further the interaction between creative processes.[21]

Yoneji Masuda, the architect of the Japanese plan for the information society, agrees with Jantsch that the primary goal of the coming age is simulated creativity. Like Jantsch, Masuda argues that a new ethics is called for—one that will reward the search for continual change and penalize behavior that questions the logic of an open-ended future.

The clock culture was accompanied by the ethic of pragmatism, a code of conduct designed to lend moral support to the idea of advancing material self-interest. The computer time culture retains pragmatism but adds to it a new ethic of simulated "creativism," a set of moral precepts designed to lend support to the idea of advancing any and all simulated activity.

In computopia, even the concept of immortality is radically transformed. Salvation is no longer to be sought in Christ on Judgment Day or in the spectre of a classless society at the end of history. In computopia, there are no final judgments or end of history. In the new world, time is information, and information is immortal. Masuda tells the next generation to put their faith in information, and salvation will be theirs. Information can be preserved against the ravages of time. It does not rot and decay; it does not get used up.

> Unlike material goods, information does not disappear by being consumed, and even more important, the value of information can be amplified indefinitely by constant additions of new information to the existing information. People will thus continue to utilize information which they and others have created even after it has been used.[22]

Astronomer Carl Sagan once mused that if we had available to us all the information stored in the genetic code of a cat, we could transmit that information to another galaxy where a superior form of being, far advanced in the science of genetic engineering, might be able to program an exact duplicate of the original. While both physical cats would eventually die, the instructions coded in the genetic information could be used over and over to program endless numbers of the same cat.[23]

In the clock culture, human beings believed that material accumulation would pave the way toward an earthly cornucopia. In the computer culture our children's generation is likely to regard information accumulation as the surefire route to everlasting salvation.

The new image of the future will profoundly affect the nature of politics. In the age of simulation, old political realities will give way to new ones at an increasingly accelerated rate. The temporal orientation of the body politic will be so sped up that there will be little time to reflect. In a political culture given over to speed and efficiency, history is more of an albatross than an anchor. The past is a weight that holds down the future, that limits options and shackles possibilities. The past places obligations on the future. Constitutions, laws, political customs, and codes of conduct have traditionally served as temporal brakes on future actions. While the political elite has always exercised control over the temporal horizon, its authority to do so has always been governed by protocols of the past. Rulers pledge to keep and fulfill past visions, promises, and commitments. The President takes an oath to protect and defend the Constitution, a historical document outlining the accumulated promises, contractual agreements, and visionary goals of the political culture. The point to emphasize is that political power has always been tempered by history. Shamans, priests, and kings have always been constrained by the past, forced to conform to historically established taboos, covenants, commandments, contracts, and constitutions.

In the age of simulation, conditions change so fast that laws, contracts, and terms of agreement become short-lived. Realities are simulated, edited, and revised at lightning speed as the body politic is tossed to and fro in a whirlpool of changing circumstances. In a political environment bombarded with novelty, there is precious little time to honor past commitments. This is because the greater the number of past commitments to be honored, the greater the amount of future time to be set aside to fulfill those obligations. Time being a premium, less and less is given over to making good on prior promises and more and more is given over to facilitating new options and agreements. As a result, political debts, like legal tender, are continually being canceled in the inflation-ridden politics of the present.

In the brave new information culture, fulfilling past political obligations will increasingly take a backseat to anticipating future political possibilities. In computopia, history is a resource to be continually reprogrammed to fit the requisites of each emergent set of realities. Memory, in the new software culture, extends little further than yesterday's information residue, which provides the data for updating tomorrow's response to today's opportunities.

Past political promises and commitments are easily toyed with and even shunted aside in a culture where "staying ahead" is the imperative and expediency the rule. In such a context, the past places fewer and fewer limits on the future. The devaluation of history is a prerequisite for the free exercise of pure power.

In order to exercise power in today's information society, leaders find they have to rely on the computer engineers and software specialists. They have become the new oracles, the anointed ones to whom we defer when it comes to securing the future.

Computer programs are designed to be predictive tools. Their creators contend that the forecasting capability of these new machines is extraordinary. It may be possible someday, in the not-too-distant future, to predict outcomes of various events with such a high degree of reliability that surprise itself will be made virtually obsolete. Jay Forrester, one of the seers of the new forecasting fraternity, laments the fact that "mental models" of the future are so "fuzzy." With the computer, says Forrester, we can begin to eliminate uncertainty for the first time in history.

> The great uncertainty with mental models is the inability to anticipate the consequences of interactions between parts of a system. This uncertainty is totally eliminated in computer models. Given a stated set of assumptions, the computer traces the resulting consequences without doubt or error.[24]

Computers already are being used to forecast in a wide range of fields that just a few short years ago relied almost exclusively on "mental models" alone. In diagnostic medicine, computers are beginning to eliminate the once secure role of the physician. Professor of computer science Edward Feigenbaum of Stanford

University boasts that computers "often outperform the very experts who have programmed them because of their medical ways; they don't skip or forget things, get tired or rushed, or fall subject to some of our other human failings."[25]

Computers are being used to predict where mineral deposits are buried deep beneath the earth's surface. They are being used to forecast global weather patterns and to model the long-term effects of petrochemicals on soil nutrients and rainwater. In the advanced technology nations, they are already an integral part of the commercial and economic life. They are being used to forecast business trends, changing market conditions, inventory needs, and consumer preferences. They are even being used in strategic military planning and, in a few years, are likely to completely replace human foresight when it comes to anticipating and responding to a potential or real nuclear attack.[26]

The new forecasting class of computer engineers and programmers enjoy greater power over the temporal horizon than any other single group in history. Joseph Weizenbaum, a pioneer in artificial intelligence, says that "no playwright, no stage director, no emperor, however powerful, has ever exercised such absolute authority to arrange a stage or a field of battle and to command such unswervingly dutiful actors or troops."[27]

In the coming computer age, power will reside with those who enjoy access to information. A top executive of a transnational corporation recently remarked that power in the boardroom is now measured by who has access to the various information codes that govern the far-flung interests of the company. "Only seven people in the entire firm," he said, "had access to all of the information codes."[28]

A new form of class exploitation is emerging at the outset of the information age. Whereas class division in the clock culture was measured in terms of management and labor, the division in the computer time culture is between those who are "told what to do by the computer" and those who "tell the computer what to do."[29]

Those who tell the computer what to do now have a tool at

their disposal with the power to monitor, manipulate, and marshal the personal lives of millions of people in ways that would have been inconceivable one hundred years ago. In his book, *Rise of the Computer State*, David Burnham, a *New York Times* reporter, reminds us that a century ago the only records kept on most Americans were birth, marriage, and death certificates; and land and home ownership records.[30] Today, sophisticated computer networks have access to some of the most intimate details of people's lives such as "whether they are seeing a psychiatrist, what drugs they use, and whether they have a drinking problem."[31] According to Burnham, there are even computer information networks that keep an account of exactly when you leave your house, when you turn on the television, when you deposit a check, and when you pick up the phone to dial a call. The computers know how much you earn in a year, how much you have in your checking account, what types of purchases you've made in the last month, what kind of vacations you like to take—even how much you like to spend on them.[32]

In the computer age, information is power and that power is becoming increasingly centralized in the hands of a small coterie of public bureaucracies and giant corporations. Burnham reports that the U.S. government alone has collected four billion "separate records" on U.S. citizens, seven items on each man, woman, and child alive.[33] "What does it mean," Burnham askes rhetorically, "when 10,000 merchants all over the country are able to obtain a summary fact-sheet about any one of 86 million individual Americans in a matter of three or four seconds from a single database in California?"[34]

The new time world of the computer age is being used most effectively by the transnational corporations, a relatively new form of institutional power which is already vying with nation-states for hegemony over territories and peoples. Today, transnational corporations do business in countries across the planet. They cross national boundaries, expropriating resources, employing vast labor pools, and exploiting consumer markets in every region of the globe. Their economic power eclipses that of most nation-states. General Motors's annual sales revenues exceeded the gross national product of Belgium and Switzerland in the 1970s.[35] Exxon's tanker fleet is 50 percent larger than the Soviet Union's.[36] Some two hundred global corporations now control nearly 80 percent of the productive assets of the non-Communist world.[37] The U.S. Chamber of Commerce predicts that within twenty-five years, these economic behemoths will own production assets in excess of four trillion dollars, or 54 percent of everything worth owning on the planet Earth.[38] In fact, thirty-six of the one hundred largest money powers in the world today are no longer countries but transnational corporations.[39] And computer technology is indispensable to their reign of power.

As Jerry Mander makes clear, "Computers not only aid today's multinational corporate enterprises, they make them possible."[40] Without sophisticated computer networks, the transnational companies would simply not be able to "keep track instantaneously with millions of pieces of information from all over the world."[41] Today's multinational corporations do business everywhere but are "not located anywhere except in the computer itself."[42] The computer is their method of communication, their timepiece, their manager, and their forecaster. The multinationals take their power with them wherever they go. It resides deep inside the microworld of silicon chips and electronic circuits that guard the data and information used to program the temporal affairs of local communities—and whole continents.

12

TIME PYRAMIDS AND TIME GHETTOS

Forecasting the future and exercising power have coexisted in a symbiotic relationship, from the very beginning of social experience until the present. As mentioned, the great leaders in history have relied on a range of predictive tools including prophecy, revelation, and science and technology to gain control over the human time horizon. They have used these forecasting skills to wile the people with extraordinary visions of the future. They have predicted the deliverance into the promised land, the coming of eternal salvation, the age of progress, and are now holding forth the spectre of a fully simulated, artificial environment on the time horizon. They have promised to lead the way to these other worlds and have called upon the faithful to sacrifice their time to the task of realizing these visions of the future.

The people, for their part, have responded. They have listened to the clarion call for a better world and have placed their time and their lives at the disposal of those who claimed to speak for the future. Some leaders have encouraged their followers to participate and share openly in the process of directing the future of the society. Still, more often than not, the great leaders have been anxious to maintain a monopoly of power over the future. They have inspired their followers to accept their view of the perfect world that awaits them, and, at the same time, have made sure to keep them totally dependent on their leadership to get there. Kurt Lewin describes the process:

164

For the distant future, to be sure, the autocratic leader fre-
quently reveals to his subjects some high, ideal goal. But when
it comes to immediate action, it is one of the accepted means
of autocratic leadership to reveal to his followers not more than
the immediate next step of his actual plans. In this way not
only is he able to keep the future of the members in his own
hands; in addition he makes the members dependent on him,
and he can direct them from moment to moment in whatever
direction he wishes.[1]

Effective rulers, then, maintain a monopoly over the knowledge
and tools necessary to predict and intervene with the future and,
by doing so, keep the people dependent on them for direction.
In ancient days, people were at the mercy of the palace priests
and the oracles. Today they are at the mercy of the "experts."

In every society in recorded history, the time hierarchy of powerless to powerful has proceeded up the same climb. On the
bottom of the pyramid are the masses, those whose time span is
narrowly circumscribed to the present. These are the people who
have been untutored in planning ahead to secure their own future. A monopoly of power in every society begins with severing
people from control over their own future, making them a prisoner
of the present. Unable to gain access to the future, people become
pawns in the hands of those on top of the temporal pyramid, who
control the human time frame.

Temporal ghettos are no less important than physical ghettos.
People confined in a narrow temporal band, unable to anticipate
and plan for their own future, are powerless to affect their political
fate. Secret police in totalitarian regimes have understood the
power of erasing people's time orientation; brainwashing techniques are designed to sever the victim from control over his
sense of time. Prisoners are subjected to continual electric light,
denied sleep, and kept away from clocks or other temporal cues.
They are stripped of recall, made to question their own memories
of past experiences. They are denied any hope for the future and
forced to live from moment to moment. Only the present exists;
the future becomes untrustworthy, unpredictable. Losing all sense

of time awareness, the victims become totally manipulable, ready to accept blindly their inquisitors' definition of past and future. They willingly obey.

Temporal deprivation is built into the time frame of every advanced society. In industrial cultures, the poor are temporally poor as well as materially poor. Indeed, time deprivation and material deprivation condition each other. A number of sociological studies conducted over the past three decades confirm the correlation between one's economic class position and one's temporal orientation. As in the past, those who are most present-oriented are swept along into the future that others have laid out for them.

In a pioneering study entitled "Time Orientation and Social Class," sociologist Lawrence L. LeShan found that the time orientation of the lower classes is much more present-oriented while the time orientation of the middle and upper-middle classes is much more future-directed. The lower classes are more apt to interact socially in terms of "quick sequences of tension and relief."[2] They are less concerned with planning and with far-off goals. According to LeShan, the lower classes perceive the future as "an indefinite, vague, diffuse region and its rewards are too uncertain to have much motivating value."[3] In contrast, the middle and upper classes tend to exhibit much longer "tension-relief sequences." They plan out their futures well in advance, then act on their plans.[4]

In the lower economic classes, the future is less predictable and economic security more problematic. Individuals find it difficult to plan ahead, as day-to-day survival is critical. Children growing up in these environments are less likely to trust the future or believe they can influence it. They narrow their horizons and goals. Children of lower-class parents tend to be more conditioned to immediate rewards and punishments. They are less likely to delay entrance into the work force in order to pursue an education in preparation for a better future. In contrast, children of the middle and upper classes tend to be more conditioned to delaying immediate gratification in expectation of larger rewards later on.[5]

LeShan tested 117 children ranging in age from eight to ten on temporal orientation. Seventy-four of the children were lower class, 43 were middle class. They were asked to tell a story and then the stories were examined "in terms of the period of time covered by the action of the story." The results confirmed LeShan's hypothesis. The middle-class children produced stories "covering a longer time span from beginning to end" than the children from lower-class families.[6]

H. Nowotny observes that the narrow time span of the poor is a logical response to the realities they face. A present-directed time orientation

> ... constitutes the only rational strategy to survive in an environment which is to a high degree uncertain, loaded with risks that are beyond the individual's control and influence, and about which only a minimum of information is available.[7]

In his book on urban poverty in America, *The Unheavenly City,* Edward Banfield likewise concluded that temporal deprivation and economic deprivation are inextricably linked:

> Extreme present-orientedness, not lack of income or wealth, is the principal cause of poverty in the sense of "the culture of poverty."[8]

The entire work life of industrial societies is saturated with temporal discrimination. Unskilled and semiskilled jobs require little past knowledge and even less predictive or planning skills. Professional jobs require both. Lacking the temporal skills that might allow advancement up the employment ladder, the menial laborer remains trapped in a series of dead-end jobs from which there is no apparent escape. Time skills limit his or her economic opportunities. Lack of economic opportunities undermine hopes for a better future and willingness to plan ahead and set long-term goals. The laborer remains stuck in a present-oriented temporal ghetto, unable to reach out and claim some measure of control over the future.

Urban anthropologist Elliot Liebow aptly summarizes the close relationship that exists between time orientation and class position:

> The future orientation of the middle-class person presumes, among other things, a "surplus" of resources to be invested in the future and a belief that the future will be sufficiently stable both to justify his investment (money in a bank, time and effort in a job, investment of himself in marriage and a family, etc.) and to permit the consumption of his investment at a time, place, and manner of his own choosing and to his greater satisfaction. But the street corner man lives in a sea of want. He does not, as a rule, have a "surplus" of resources, either economic or psychological. Gratification of hunger and the desire for simple creature comforts cannot be long deferred. . . . Living on the edge of both economic and psychological subsistence, the street corner man is obliged to expend all his resources on maintaining himself from moment to moment.[9]

Liebow identifies "surplus" as a key factor in determining where people line up on the temporal hierarchy. Surplus or stored wealth is one of the revolutionary concepts in the history of our species and helps define the temporal pyramid in every civilization. For 99 percent of our existence, the human race lived free of the idea of surplus. Over the long stretches of Paleolithic time, hunter-gatherers existed day to day, subject to the vicissitudes of nature, never able to secure a safe tomorrow. The very idea of storage had to wait for the development of a novel technology: containers.

It was not until the early Neolithic era that women began to experiment with crude pottery, providing containers to store grain and other foodstuffs. Stored surplus allowed human beings to free themselves from the periodicities of the environment. People could for the first time plan ahead by storing grains, ensuring against the prospect of drought, flood, or other calamities.[10]

Surplus generated not only wealth but also a spate of unresolved questions. How much surplus should be stored, who should decide, who should control the storage, who should distribute the surplus, and how should it be distributed? Thus, politics began with the first piece of pottery.

Surplus allowed human beings the first opportunity to dictate the terms of the future. Managing surplus wealth and securing the future became one and the same thing. Those in control of the stored wealth invariably controlled the future.

If ever there was a need to confirm the importance of stored wealth in the exercise of temporal power, the industrial age provides the ideal case study. The raw resource of the industrial age is surplus sun, in the form of long-buried deposits of coal, oil, and natural gas. For the first time in history, human beings could free themselves from total reliance on the energy provided by the sun's rays. Dipping down into the burial grounds of the Carboniferous period, humanity expropriated and harnessed millions of years of stored sun, breaking loose from the age-old dependency on solar rhythms.

With stored sun at our disposal, it became possible to superimpose a new set of accelerated mechanical rhythms onto the natural rhythms of plant and animal life and the social rhythms of human life. This new form of concentrated wealth gave humanity the energy surplus it needed to speed up the life cycle of both nature and culture to ever more exacting standards of efficiency.

The individuals, groups, classes, and institutions that controlled this new form of stored wealth dictated the terms of the new, industrial landscape. They decided how the time horizon was to be filled in by exercising power over how the new wealth was to be used. The new fossil-fuel culture was accompanied by a host of new economic and social forms including the introduction of automated machine technology, mass transport, the development of a mobile work force, and urbanized living patterns. These radical innovations dramatically altered the temporal orientation of much of the human family.

The nation-state, the corporation, the merchant class, the bourgeois entrepreneur, the scientists, and the technicians emerged as the agents of the new time frame. They managed the new stored wealth and, with it, the future horizon and human time dimension as well.

Now, as we begin to make the transition out of the industrial age, information becomes the new form of stored wealth. In the computer age, temporal power will be defined as having access to and control over raw data extracted from past experience, and being able to effectively program future realities. Already the battle for control of this new form of wealth is generating tension between various groups in society. Unions, citizens associations, and civil libertarians, all of whom are affected by the computer revolution, are demanding access to and eventual control over stored information that could determine the future of society.

Access to such information has even caught the attention of highly educated computer criminals who are learning how to break into the most sophisticated computer programs in order to manipulate or steal vital information for commercial gain. Computer piracy is now a major concern of the information industry. It would seem that whoever controls the data and software of the nanosecond culture will determine how the future will be programmed. Indeed, the battle lines are being drawn between those who believe in "enriched informational skills" on the one hand and those who are computer illiterates on the other, conjuring up an image of new class boundaries separating the powerful from the powerless, the privileged from the exploited.[11]

PART IV

COSMIC TIMEPIECES AND POLITICAL LEGITIMACY

13

THE CLOCKWORK
UNIVERSE

We have described how societies divide up the time of their members with time-allocating devices like calendars, work schedules, and computer programs. The people willingly give up control over their own time in return for the promise by their leaders that an idyllic, timeless future will be their reward. As we will now see, those in power claim cosmic legitimacy for the way they manipulate and exploit the time of their subjects by formulating a view of nature congenial to the way they organize and control the society's temporal affairs. In the industrial age, Western cosmologists and philosophers painted the picture of a "clockwork universe," a view of nature that mirrored the key operating assumptions of the clock and schedule culture. As we move into the computer age, a new generation of intellectuals is redefining the cosmos as an "information universe," a reality that comports quite well with the assumptions of the emerging nanosecond time frame. Cosmologies, then, serve as the ultimate form of justification for our day-to-day temporal activity. They allow the social order to continue the fiction that its behavior conforms "with the natural order of things."

In 1377, a French scientist and philosopher, Nicole d'Oresme, coined the term "the clockwork universe."[1] Oresme looked to the heavens and saw the same principles at work as could be found in the workings of a master clock. Over the next several centuries, European scholars increasingly relied on the clock metaphor to explain the workings of the universe, nature, living systems, and the body politic. The clock, which was reordering the time frame and the temporal consciousness of much of Europe, was also being looked to as a convenient model for reordering the metaphysics of existence. Thus, God became the first casualty of the new clockwork universe.

At the height of the Church's reign over medieval Europe, St. Thomas Aquinas likened God to a craftsman, an artist who had expertly crafted his creation and then continued to work along with his appointed emissaries on earth to perfect his handiwork. The modern scholars provided God with a new vocation. He became the Master Clockmaker, who had assembled the universe by the same sort of mechanistic principles used to assemble a clock. In place of the personal God who intervened into the affairs of the world and actually took part in the life of his creation, the clockmaker God was remote and, if not uncaring, at least impartial.

There was simply no room in this orderly world for an arbitrary or capricious God, or even a God who might decide to change the rules from time to time. Just as a beautifully designed clock runs by a prearranged, prefixed set of principles and rules, so, too, does the clockwork universe and everything in it. God, it was suggested, would no more tamper with the operating mechanism of the universe than a clockmaker would his own finely tuned timepiece.

Indeed, one of the most hotly contested questions of the period was whether God needed to intervene at all after having wound up the cosmic clock at the dawn of time. Gottfried Wilhelm Leibniz took issue with Isaac Newton over just this question, chiding the great mathematician for constructing a theory of physical forces and time that required God to rewind the cosmic clock from time to time. Surely, remarked Leibniz, God, in his infinite foresight and wisdom, would have made sure to incorporate the idea of "perpetual motion" into his universal clock.[2]

Leibniz was not about to open the door for a return of God into the affairs of the world. Accordingly, he said that it would be unfitting even to suggest that God might have to readjust the universal clock. As God was the perfect clockmaker, it only made sense that "his machine lasts longer, and moves more regularly, than those of any other artist whatsoever."[3] The only logical conclusion one can come to, said Leibniz, was that "the wisdom of God consists in framing originally the perfect and complete idea of a work, which began and continues . . . by the continual uninterrupted exercise of his power and government."[4]

Having cast God in the role of cosmic clockmaker, European philosophers were understandably anxious to redefine his earthly creation using clocklike metaphors. Like God, nature fell victim to the new temporal reductionism. In *Novum Organum*, Francis Bacon compares the workings of a clock to the workings of the planets and the life force of animals and concludes that "the making of clocks . . . is certainly a subtle and exact work; their wheels seem to imitate the celestial orbs, and their alternating with orderly motion, the pulse of animals. . . ."[5]

Descartes followed suit, likening animals to "soulless automata" whose movements were little different from those of the automated puppetry that danced upon the Strasbourg clock. When an animal whimpers or screams, Descartes declared, it is not showing feeling or pain but only emitting sounds and noises from its internal gears and mechanisms. "It seems reasonable," Descartes suggested, that "since art copies Nature, and men can make various automata which move without thought, that Nature should produce its own automata, much more splendid than artificial ones. These natural automata are the animals."[6]

The clock metaphor captured the imagination of the best minds in Europe. Philosophers and poets, scientists, and theologians projected the language of the clock onto the universe. Just as the clock was becoming instrumental in reshaping the social and economic order, it was proving to be equally important in revising our understanding of the totality of existence.

While others pondered the clock metaphor, Isaac Newton transformed the analogy into a synonym. His laws of matter and motion were presented to the world as proof positive that the same "universal laws that governed the smallest portable watch also governed the movements of the earth, the sun, and the planets."[7] Newton became the celebrated author of a new concept of time. He confirmed what his predecessors had already suspected: Time is an objective phenomenon, independent of events that human beings may care to attach to it. According to Newton, "Absolute, true and mathematical time of itself, and from its own nature, flows equably without relation to anything external."[8]

It has been said that science owes more to the steam engine than the steam engine owes to science. The same might be said of the clock. Even before Newton had set forth the new science of time in his *Principia,* the concept had already been partially absorbed into the daily ritual. Men and women were beginning to gaze upon the clock as a kind of alien force, a detached, impartial overseer that metes out time with relentless and exacting regularity, seemingly unaffected by the unfolding of either natural phenomena or social events.

Newton and his contemporaries insulated clock time from the inconsistencies and incongruities of social life. It was an independent force that no mere mortal could affect, even as it affected the life of every human being. This new form of temporal power was impressive. Imagine a force that regulated every nuance of reality, a force so potent and inclusive that it could rule over the planets and stars and, at the same time, regulate even the lowliest affairs of plants, animals, and man.

If the architects of the modern age were in search of order, they secured their end and then some. In the new clockwork universe, every single thing was ordered to meet the rigorous standards of the new mechanical time. Nothing was left to chance. Everything was made safe and secure and, above all, predictable.

Even the assumptions underlying political philosophy and economic theory were made to conform with the new clockwork metaphor. The great political philosopher John Locke argued that the same clocklike principles that comprised the natural laws of the universe were exactly those that ought to regulate the affairs of human beings on earth. The Scottish economist Adam Smith observed that, just as a pendulum regulates the proper functioning of a clock, the invisible hand of nature operates in similar fashion, assuring the proper regulation of supply and demand in the economic marketplace.

The philosophers of the modern era recast their image of the universe, nature, and human affairs in a way that served to legitimize clock culture. Their new image of God and reality could not have been more warmly received by the new class of merchants, entrepreneurs, and factory owners. They could henceforth justify the new time conception they were imposing on the culture by arguing that it was merely a reflection of "the natural order of things." The clock, then, was used to conjure up the idea of a clockwork universe, and the clockwork universe, in turn, was used to justify the idea of using the clock to regulate the social and economic order. The new clock culture became imprisoned in its own tautological jail cell.

In just a few short centuries, the bourgeois class had managed to hoist the mechanical clock to the top of the town tower and then succeed in lifting its spirit up into the heavens where, like the angel Gabriel, it proclaimed the coming of the kingdom. The promised land, however, bore a strikingly secular imprint. God's countenance, which once shone brightly, now cast only a pale shadow. The sounds of divine rapture could no longer be heard. They were subsumed by the relentless ticking of the giant cosmic clock. Underneath its watchful gaze, the faithful scurried to and fro, frantic to keep up with the tempo of the times, anxious not to miss a single beat for fear that they might be forever condemned to that netherworld where no clocks existed and mayhem and confusion ruled supreme.

The clockwork universe was not without its fair share of critics. There were those who, for religious, aesthetic, and economic reasons, were less than anxious to subscribe to this new order in either its philosophical guise or secular form. In fact, the clock had barely made its debut on the world scene when, in 1340, an early critic offered a scathing review of its domineering presence:

> Confusion to the black-faced clock by the side of the bank that awoke me! May its head, its tongue, its pair of ropes, and its wheels moulder; likewise its weights of dullard balls, its orifices, its hammer, its ducks quacking as if anticipating day and its ever restless works. This turbulent clock clacks ridiculous sounds like a drunken cobbler. . . . The yelping of a dog echoed in a pan! The ceaseless chatter of a cloister. A gloomy mill grinding away the night![9]

If the clock was a source of irritation, the cosmology that accompanied it was greeted with even greater disdain. Some, like the Jesuit father Guillaume Bougeant, were incredulous that Descartes and his ilk were attempting to reduce all of nature, even animals, to clocklike principles:

> I defy all the Cartesians in the World to persuade you that your Bitch is a mere machine. . . . Imagine to yourself a Man who should love his watch as we love a Dog, and caresses it because he should think himself dearly beloved by it, so as to think when it points to Twelve or One O'clock, it does knowingly and out of tenderness to him.[10]

Jonathan Swift took on the new clockwork cosmology with biting satire in *Gulliver's Travels*. He describes the reaction of the Lilliputians upon first seeing Gulliver's watch:

> Out of the right fob hung a great silver chain with a wonderful kind of Engine at the Bottom. . . . He put this Engine to our ears, which made an incessant Noise like that of a Water Mill. And we conjecture that it is either some unknown Animal, or the God he worships. . . . But we are more inclined to the latter opinion, because he assured us that he seldom did anything

without consulting it. He called it his Oracle, and said it pointed
out the time for every action of his life.[11]

Aware that the clock metaphor was being increasingly used
by politicians to define, explain, and justify the machinations of
state power and political intrigue, poet Alexander Pope speculated
that "perhaps it may be with states as with clocks, which must
have some dead weight hanging at them to help regulate the
motion of the finer and more useful parts."[12]

14

THE INFORMATION
UNIVERSE

The clockwork universe of Bacon, Descartes, Newton, and Locke is being challenged by a new schema whose basic assumptions are borrowed directly from the operating principles of the new computer technology. A new cosmology looms on the horizon and with it a new language for interpreting reality. Today, cosmologists no longer speak of "winding up" the cosmic clock, "assembling" reality, or "balancing" forces. They speak, rather, of "programming" the universal computer, "processing" phenomena, and maintaining negative and positive feedback loops. The clockwork universe of the modern era is being subsumed by the new computerlike universe. Philosophers, scientists, and theologians are consecrating our newest technology with the highest honor that can be bestowed on a human artifact. They are re-thinking and remaking the world to conform with the essential attributes of this newest contrivance.

A powerful new form of reductionism is spreading through the intellectual community, capturing the attention and igniting the imagination of a whole generation of thought makers. The old mechanical clockwork reductionism that provided the gist for centuries of inspired thinking about the nature of Nature appears to have run its course. In its place intellectuals talk excitedly and expectantly about the "computerlike" way all physical and biological phenomena seem to behave. This new view of nature adds a veneer of legitimacy to the new technology as all of reality is seen as operating by a set of governing principles remarkably similar to those that govern the computer. The set of theoretical principles that underlies computer technology is known as information theory. It is these principles that are now being used to redefine our view of the world.

Information theory was the brainchild of Claude Shannon of Bell Laboratories. In 1948, Shannon published two papers that contained a set of theorems dealing with how messages are sent from one place to another in the most efficient manner. The intellectual community beat a path to the new theory, comparing its impact to that of Newton's laws of motion. Shannon's theory, which was to become the scientific basis for the nascent computer technology, was seized upon with great enthusiasm by a range of disciplines because its assumptions could be so readily applied to such a wide spectrum of phenomena. Here was a new ordering principle with practical as well as metaphysical significance. As a set of theorems, it provided fertile technological soil for the development of the new computer technology. As a scientific concept, it provided humanity with a new lens for viewing reality.[1]

Information theory has already been used to redefine fields as diverse as astronomy, biology, and psychology. Astronomers, like David Layzer of Harvard, now view the expanding universe as an ever-evolving system of information. According to Layzer, the universe is always generating new bits of novelties, islands of information that grow more complex with the passage of time. Planets, stars, and galaxies represent rich islands of information that continue to gain in complexity at the expense of an increase in entropy in the universe as a whole. The old view of the universe pictured the cosmic clock slowly winding down toward equilibrium and heat death. The new view pictures the universe as a field in which energy is consumed, generating increasing information. In the old clockwork universe, time was set at the very beginning of creation and then ticked away in a linear fashion until the coil loosened and the gears gave out. In the new information universe, time is information and continues to expand. Information begets information; complexity begets complexity in an ever more structured cosmos.[2]

The new universe, then, operates more like a computer than a clock, with information generating new information at every stage and with complexity becoming increasingly dense and structured.

To appreciate how information generates order and increasing complexity, it is necessary to understand a second theoretical concept that underlies computer technology. Cybernetics emerged on the heels of the information theory, providing the explanation of how information could continue to build up even as the universe itself continues to run down. Norbert Wiener, the father of cybernetic theory, began by defining information as the countless messages that go back and forth between things and their environment. Cybernetics, in turn, is the theory of the way those messages or pieces of information interact with one another to produce predictable forms of behavior.

According to cybernetic theory, the steering mechanism that regulates all behavior is feedback. Consider the thermostat, for example. The thermostat regulates the room temperature by monitoring the change in temperature in the room. If the room cools off and the temperature dips below the mark set on the dial, the thermostat kicks on the furnace, and the furnace remains on until the room temperature coincides once again with the temperature set on the dial. Then the thermostat kicks off the furnace, until the room temperature drops again, requiring additional heat. This is an example of negative feedback. All systems maintain themselves by the use of negative feedback. Its opposite, positive feedback, produces results of a very different kind. In positive feedback, a change in activity feeds on itself, reinforcing and intensifying the process rather than readjusting and dampening it. For example, a sore throat causes a person to cough, and the coughing, in turn, exacerbates the sore throat.

Cybernetics is primarily concerned with negative feedback. As the examples illustrate, feedback provides information to the machine on its actual performance, which is then measured against the expected performance. The information allows the machine to adjust its activity accordingly in order to close the gap between what is expected of it and how it in fact behaves. Cybernetics is the theory of how machines self-regulate in changing environments. More than that, cybernetics is the theory that explains "purposeful behavior" in machines. For Wiener, all purposeful behavior reduces itself to "information processing."

> It becomes plausible that information . . . belongs among the great concepts of science such as matter, energy, and electric charge. Our adjustment to the world around us depends upon the informational windows that our senses provide.[3]

With information theory and cybernetics, technologists were armed with the scientific principles they needed to design the first generation of automated computers, machines that could interact with the environment and adjust to it in a self-regulating mode. Wiener and his colleagues, however, were not satisfied to limit their new theories to machines. Convinced they had unlocked the principles underlying purposeful behavior in the universe at large, the information theorists of the early 1950s set out to extend their theorems to every aspect of reality, including living systems.

It was Wiener's thesis that "the physical functioning of the living individual and the operation of some of the newer communication machines are precisely parallel."[4] Wiener compared the new automatic machines to the nervous system in animals and concluded that they are "fundamentally alike." Living things and machines both collect and process information from their environment and generate new information that allows them to adjust to changing conditions in the outside world.

In the past thirty years, information theory and cybernetics have treaded directly into the life sciences, effectively redefining biological phenomena to suit the vision first laid out by men like Shannon and Wiener. The principles that have given rise to the computer revolution have been so thoroughly integrated into the world of biology that computers and living things now share a common language.

In their book, *Current Problems in Animal Behavior,* psychologist Oliver Zangwill and zoologist William Thorpe of Cambridge University observe that the "principles derived from control and communication engineering are being increasingly brought to bear upon biological problems, and 'models' derived from these principles are proving fertile in the explanation of behavior."[5] The fact is, scientists in the biological field are redefining much of the working language of their discipline to conform with the underlying principles of information theory and cybernetics. William Thorpe and other biologists now look upon living things as systems that "absorb and store information, change their behavior as a result of that information, and . . . have special organs for detecting, sorting, and organizing this information."[6] Like many of his colleagues, Thorpe borrows the idea of computer programs to explain how the life process works. "The most important biological discovery of recent years," says the zoologist, "is that the processes of life are directed by programs . . . and that life is not merely programmed activity, but self-programmed activity."[7]

Where scientists and philosophers of the Enlightenment projected the principles of the mechanical clock onto the blueprint of nature, a new generation of scientists and philosophers is projecting the principles of the new computer technology onto living things. Today life is viewed as a code containing millions of bits of information capable of being programmed in myriad ways.

Philosopher Kenneth Sayre of Notre Dame summarizes the newfound consensus when he states that "the fundamental category of life is information."[8] The world-renowned French biologist Pierre Grassé goes even further, suggesting that all life can be redefined in cybernetic terms. Grassé compares the operating principles of the computer and computer programming to the way the genetic code of an organism works and concludes that both are governed by the principles of information theory and cybernetics.

Grassé's use of metaphor is not surprising. It should be remembered that even at the very beginning of the genetic age, biophysicists James Watson and Francis Crick explained their discovery of the DNA double helix using the new information language. The gene, they announced, is a code containing many bits of information programmed into specific sequences.

Information theory and cybernetics are also being used to modify evolution theory. The old Darwinian view of "survival of the fittest" is now being cast aside in some quarters in favor of a new view of "survival of the best informed." According to standard Neo-Darwinian orthodoxy, each species up the evolutionary line is better able to utilize scarce resources more efficiently. The emergent theory characterizes each species up the evolutionary chain as better adept at processing greater stores of information in shorter time spans. In the clockwork universe, organisms were perceived of as "expertly built machines." In the new information universe, organisms are thought of as "well-designed programs."[9]

In the old industrial time world, philosophers like Adam Smith and John Locke contended that every person sought to maximize his or her own material well-being and by doing so, inadvertently enhanced the common good. Darwin lent an air of scientific legitimacy to such claims by finding a similar pattern in nature. Even though each organism fights for its own self-sufficiency, it inadvertently advances the common good by contributing its improved traits to the evolutionary process.

In the information universe, the idea of maximizing self-interest is replaced with the idea of maximizing self-organization. According to the new schema, every organism seeks to maximize its own self-organization by exchanging information with its environment. While each species seeks only its self-organization, in the process it generates new bits of information that are the source of further evolutionary development. Every new evolutionary advance, in turn, increases the overall complexity of the system, further integrating all the information into a richer labyrinth of relationships.

This cybernetic explanation of evolution provides humanity with an updated rationale for its continued manipulation of the environment. In an age steeped in the information mystique, people can take comfort from the belief that their own efforts to generate and control greater stores of information not only advance their self-organization but also contribute to the strengthening of all relationships in nature by increasing the level of interaction, interconnection, and synchronization in the system as a whole.

Information theory is even being used as a rationale for why human beings occupy a privileged position in the evolutionary scheme of things.

> Human beings . . . excel in the acquisition of information, and also in versatility of information processing . . . since superiority in information gathering and processing amounts to superior adaptive capacities, this accounts for human dominance over other kinds.[10]

While the principles underlying computer technology have influenced much of our thinking about the world around us, their biggest impact has been on how we conceive of the nature and function of the human psyche. Information theory and cybernetics have run roughshod through the field of psychology, converting much of that discipline into a way station for the latest ideas in computational thinking.

> Many psychologists have come to take for granted in recent years . . . that men and computers are merely two different species of a more abstract genus called information processing systems.[11]

Some psychologists have been so enthusiastic in their response to the new technology that they have given birth to a new field in recent years. Cognitive psychology brings together computer science, information theory, and psychology into a unified realm of study.

In the cognitive model of the human mind, information is received by the system, held in buffer stores or permanent memory, processed according to intelligent routine, and used to initiate behavioral and other responses.[12]

Sherry Turkle provides some insight into why the new field of cognitive psychology has been so effective in winning over new converts. She finds that while traditional Freudian psychology seemed "speculative to some, literary to others, the computational model arrives with the authoritative voice of science behind it."[13] The real danger in this new marriage between the computer sciences and psychology rests in its perception of psyche. It used to be that people thought of the computer in human terms. Increasingly, however, people are beginning to think of themselves "in computational terms."[14]

Turkle spent several years interviewing MIT and Harvard students on the subject of human intelligence versus computer intelligence and found that "the idea of thinking of the self as a set of computer programs is widespread."[15] Many of the students compared the human brain to computer hardware and the human mind to software. The brain contains neurons, synapses, electrical impulses in the same way the computer contains electronic circuits and bits. The mind forms images and thoughts the same way a program processes information into instructions for action.[16]

Interestingly enough, many of those who have come to express this new way of thinking about the nature of thought see no room for free will in their models. Mark, a student at MIT, when asked about the concept of free will, was unequivocal in his conviction that any such notion was pure illusion:

> You think you're making a decision, but are you really? For instance, when you have a creative idea, what happens? All of a sudden, you think of something, right? Wrong. You didn't think of it. It just filtered through. . . . A creative idea just means that one of the processors made a link between two unassociated things because he thought they were related.[17]

When asked if the human mind "is anything more than the feeling of having one," Mark replied in a fashion that seems to capture the entire temporal consciousness of the first generation of compulsive computer users. He said, "You have to stop talking about your mind as though it were thinking. It's not. It's just doing."[18]

As to where emotions fit into this new cognitive model, many would agree with Masanoa Toda, a respected Japanese psychologist who contends that "emotions are decision-making programs developed through evolution."[19] When Mark was asked about emotions, he said, "Okay. Let's model it on a piece of paper."[20]

The nanosecond culture brings with it a new and more virulent form of reductionism. The clockwork universe of the industrial age is being replaced, in fast order, by the computational universe of the postindustrial age. For several hundred years Western culture has defined mind and matter in mechanistic terms, reducing all of reality to the operating principles of clockwork technology. Now, a new journey begins. In the coming century our children are likely to redefine their environment using the language of information theory and cybernetics as they attempt to conjure up a view of nature that conforms with the operational principles of the new computer technology. We are entering a new temporal world where time is segmented into nanoseconds, the future is programmed in advance, nature is reconceived as bits of coded information, and paradise is viewed as a fully simulated, artificial environment.

PART V

THE DEMOC-
RATIZATION
OF TIME

15

TEMPORAL TREKS
AND FUTURE OPTIONS

In the last century, the human race has passed into the fourth great era of time consciousness. Each era has been accompanied by its own unique temporal reality.

We have ordered our lives and our times by the seasons, the stars, the clocks, and now computers. Each new time "reckoning" system has been accompanied by a new time "ordering" system. We have used rituals, calendars, schedules, and programs to bind the human species into communities of living.

As we have moved from one temporal era to another, our vision of meanings and ends, our understanding of *eidos* and *telos*, has taken on new shapes and forms. Our early ancestors coveted the circle, perceiving time as eternal return, a ceaseless repetition of an endless cycle of birth, life, death, and rebirth. Later, as ritual consciousness gave way to religious consciousness, the vertical line of spiritual ascent replaced the circle in the Western portion of the globe as men and women looked skyward for their temporal inspiration. During the short reign of historical consciousness, the horizontal line of progress ruled as the undisputed signature of the period. Now, still in the early decades of psychological consciousness, it is the spiral that commands our attention. It is the new symbol of creation captured in both the double helix and in the cybernetic vision where feedback loops simulate new worlds pulsing in the crevices of millions of silicon chips.

With each new time-reckoning and time-ordering system, humanity has distanced itself farther and farther from the rhythms of nature. From participatory union to astronomical oversight to mechanical intervention to electrical simulation, the temporal trail leads away from the intimacy of shared temporality that binds life to life, human to beast, animal to plant. We used to perceive time as being imbedded in natural events. Now we perceive it as an external symbol, a quantified abstraction.

The living earth, whose once-powerful rhythms resonated with clarity, now appears frail and weary in the presence of a field of artificial rhythms crisscrossing the planet and beaming up into the cosmic reaches.

The human race has traveled across many different time zones in its short history on earth. With each new journey, we have been able to gain greater insight into the workings of the temporal design. We have been able to step back and gaze over vast stretches of time to witness the many connections and causations that punctuate the time world of nature. Our improved memory skills and forecasting abilities have given us a deep understanding of the time dimensions of the biological and physical realms we inhabit.

While revealing, we have paid dearly for the knowledge we have gained. We have come to see how things fit together by separating ourselves from nature's biological clocks. Our vision has provided us with great detail, but the distance we set between ourselves and the rest of creation has left us far removed from the rhythms of intimate temporal participation. We gained perspective, and in the process we lost touch with the ground of our temporal being. Our knowledge has been our alienation. Our increased understanding of nature has been accompanied by a self-imposed exile from biological time.

At each new temporal crossroads of our existence, we have made a conscious choice to use our increased perspective to secure increased power. We have sacrificed wisdom for violence and used awareness as a weapon to secure our temporal domination.

The joining of perspective to power has caused us to view the world in the most narrow of terms. The fact is, we have chosen to sever our participatory union with the rest of creation, and to redefine the world as a binary field where only subjects and objects exist. We have chosen autonomy over participation, isolation over communion, and have used power to turn the world's phenomena into objects for manipulation and expropriation. Our increasing temporal awareness has been put to the service of mastery. We have used our temporal skills of reflection and anticipation, hindsight and foresight, as time traps to snare and exploit the rhythms of the life world and to subdue and enslave our fellow human beings. We have tilted the time world of the planet to our advantage, coercing the rhythms of nature into conformity with our own social sense of sequencing, duration, tempo, and timing.

It is ironic that we have done all of this in the name of security. We have attempted to use time to secure our future and in the end each of our tomorrows seems less certain and more bleak. Contemporary Western culture has been obsessed by the idea of saving time and extending duration. Yet, we appear to have left our children with less and less of a future to enjoy. Nuclear warheads bring us within minutes of total annihilation. Air, water, and land are polluted by acid rain, chemical and nuclear waste dumps, and thousands of petrochemical compounds that cannot be recycled.

The high-tech world of clocks and schedules, computers and programs, was supposed to free us from a life of toil and deprivation, yet with each passing day the human race becomes more enslaved, exploited, and victimized. Millions starve while a few live in splendor. The human race remains divided from itself and severed from the natural world that is its primordial community.

The alliance of perspective and power has been formidable. We now orchestrate an artificial time world, zipping along the electronic circuits of silicon chips, a time world utterly alien from the time a fruit takes to ripen, or a tide takes to recede. We have sped ourselves out of the time world of nature and into a fabricated time world where experience can only be simulated but no longer savored. Our weekly routines and work lives are punctuated with artificial rhythms, the unholy union of perspective and power. And with each new electric dawn and dusk, we grow further apart from each other, more isolated and alone, more in control and less self-assured.

So anxious have we become in this increasingly frenetic and fleeting time world of clocks and computers, schedules and programs, that we often count the days to our next vacation and the prospects of escaping to the quiet calm of a forest glen, an ocean beach, a rolling meadow, anywhere far away from the incessant clamor of mechanical beats and electronic buzzes. Holidays seem to offer the only respite from the accelerated time demands of the modern world. On our weekends and weeks off, we make a hasty retreat away from the rhythms of culture and back to the rhythms of nature in an attempt to reconnect with the origins of our existence: the organic time world in which we once dwelled.

Our experts tell us that our attempts to rejoin the natural world are futile, that we can never experience nature the way our ancient ancestors once did. The human psyche is too well developed to revert back to a preconscious existence. The experts tell us that we can no more erase the history of our consciousness than we can reverse the time of our journey.

While it is true that we can no longer go back to an undifferentiated state of nature, we can make a choice to go forward to a new partnership with the rest of the living kingdom—a partnership based on a deep and abiding respect for the rhythms of the planet. To do so, we would have to be willing to give up the long-standing alliance of perspective and power and seek a new temporal orientation based on an empathetic union with the biological and physical clocks of nature.

Before us lies the great challenge of evolving human time

consciousness: to finally separate perspective from power and unite the former in a new alliance with empathy.

In the new alliance we use perspective to develop an empathetic appreciation for and commitment to the kingdom of life. Empathy heals the gaping wounds occasioned by the objectification of the world. Empathy is a path that leads from perspective to intimate participation. In all empathetic journeys, the "I" travels beyond itself, leaving itself behind as it enters into the soul of the other. To empathize is to lose oneself in the other, to be willing to surrender autonomy and identity, to give up isolation and separation in favor of becoming part of an organic community.

We have all experienced rapture, those special moments of undefinable time when we surrender control over the future, both our own and others'. We become our environment. Our time expands to encompass all other things; we become the world. We participate fully in a shared temporality where there exists no time hierarchy, no privileged minority, no inside and outside, no "I" and "they." These are the moments of bliss we yearn for and that can only be attained by empathetic, shared participation.

The union of perspective and empathy is not a new experience. Most people through history have lived in both worlds of time consciousness, the one coupling perspective with empathy, the other coupling perspective with power. While both forms of consciousness have existed from the beginning of human culture, history shows that with only scattered exceptions, empathetic consciousness has been largely relegated to the private world, whereas power consciousness has dominated the public life of civilization. Over and over again, perspective has been used to exert power over the future. Time politics has been power politics.

By transferring our private experience of empathy into public policy, we begin a new time journey, one in which temporal awareness is used to empathize with the future. In this new temporal world, time politics becomes empathetic politics. Hindsight and foresight are no longer used to gain power over the future or to impose accelerated, artificial rhythms onto the natural time world. On this new time plateau, we use our consciousness to gain perspective of how nature and life unfold over time. We become more sensitive to the workings of ecological and cultural succession. We come to perceive life not in terms of frozen future segments to be manipulated but as an unending continuum that requires stewardship and demands respect.

In an empathetic time world, there is less likelihood of exploitation. Bondage does not easily flourish in a world where time is truly shared.

In an empathetic political order, the most inalienable of all rights is the right to share equally in time. Everyone has a right of equal access to both past and future. In an empathetic time world, planning the future is a communal venture and memorializing the past a shared undertaking. There is no thought of possessing time, much less of manipulating the time of others to secure an advantage over the future. Robbing people of their past and future is an unthinkable crime in a social order where empathy is the ruling paradigm.

The democratization of time becomes the overriding priority in a society where empathy substitutes for power. Temporal pyramids are eschewed and hierarchical schedules and programs are replaced with shared-time tasks as people come to view time as a collective experience rather than a tool to exercise power over others.

In a hierarchical time culture, status is often delineated in terms of how valuable a person's time is. The time poor are made to wait, while the temporally privileged are waited upon. Material compensation is less determined by work accomplished than by the notion that some people's time is more valuable than others' and therefore worthy of greater remuneration.

Political tyranny in every culture begins by devaluing the time of others. Indeed, the exploitation of human beings is only possible in pyramidal time cultures, where rulership is always based on the proposition that some people's time is more valuable and other people's time more expendable. In a democratic time culture, everyone's time is valuable and no one's time is any more expendable than another's. In a culture where the sacredness of all life comes first, there can be no other way to view time.

The concept that everyone's time is equally valuable is truly revolutionary. The democratization of time leads to a very different social order, one in which time priorities and restraints are equitably shouldered. In such a society, everyone treats other people's time with the same regard as their own. The revaluation of time is a prerequisite to the revaluation of life.

16

BEYOND LEFT AND RIGHT

While time has permeated and saturated the political arts from the very beginning, its pivotal role in the affairs of society has gone unexplored and largely unexamined. Up to now temporal awareness existed just below the surface of recognition, always influencing and shaping the experience of our species, but never given much overt attention as a key force in the historical process. Now that time awareness has risen to the surface of our collective consciousness, it is beginning to offer a variety of new metaphoric opportunities for rethinking and reimaging the political process.

It is likely that in the years ahead, time considerations will play a more direct role in determining the contours of political action. Nowhere will their influence be more keenly felt than in redefining political identifications. Partisan loyalties have long been steeped in spatial metaphors. We need look no farther than the traditional markers used to denote one's political persuasion in most societies: right wing and left wing. Many find it increasingly difficult to identify with either "side" of the prevailing political spectrum. In a world that is becoming ever more aware of the temporal underpinnings of political action, the long-standing spatial definition of political allegiance is slowly and inexorably losing its metaphoric appeal.

It is possible, then, that the spatial alignment of right to left will be challenged by a new temporal alignment as increasing awareness of temporality pervades our conceptual vocabulary. The vague outlines of such an alignment are beginning to impress themselves onto the political process. The new time spectrum runs from empathetic rhythms on one side to power rhythms on the other. Those who align themselves with the empathetic time dynamic are calling for the "resacralization" of life at every level of existence from microbe to man. Those aligning themselves with the power time dynamic are calling for a more efficient simulated environment to secure the general well-being of society. The rhythm of the first constituency is slow-paced, rhapsodic, spontaneous, vulnerable, and participatory. Emphasis is on reestablishing a temporal communion with the natural biological and physical rhythms and of coexisting in harmony with the cycles, seasons, and periodicities of the larger earth organism. The rhythms of the other side are accelerated, predictable, and expedient. Emphasis is on subsuming the natural biological and physical rhythms and creating an artificially controlled environment that can assure an ever-increasing economic growth curve for present and future generations.

Each of the new temporal poles offers a radically different sense of time—both of them alluring, each equally appealing. In the final analysis, however, these two time orientations are utterly incompatible. While they may be able to coexist for a time in a political world that is comfortable with the notion of absorbing contradictions and negotiating compromises, in the end one of these two temporal orientations will ultimately prevail, providing a context for a new vision of the social order.

The empathetic temporal orientation gives rise to an ecological, stewardship vision of the future. Its advocates would like to establish a new partnership with the rest of the living kingdom. At the heart of this new covenant vision is a commitment to develop an economic and technological infrastructure that is compatible with the sequences, durations, rhythms, and synergistic relationships that punctuate the natural production and recycling activities of the earth's ecosystems. Proponents believe that social and economic tempos must be reintegrated with the natural tempos of the environment if the ecosystem is to heal itself and become a vibrant, living organism once again.[1]

The power temporal orientation gives rise to a high-technology simulated vision of the future. In this time world, an ever more complex and sophisticated labyrinth of fabricated rhythms will increasingly replace our long-standing reliance and dependency on the slower rhythms of the natural environment. Advocates of the power temporal orientation envision an artificial environment regulated by the sequences, durations, rhythms, and synergistic interactions of computers, robotics, genetic engineering, and space technologies; an environment where order, foresight, predictability, and efficiency have replaced the uncertainties and anxieties that have plagued the human family since the dawn of civilization.

Advocates of power time believe that security is achieved through control over the temporality of nature. Advocates of empathetic time believe that security is achieved by participating in communion with the pulse of the larger communities that make up the ecosystems of the planet. The sharp contrast in these two different temporal and political visions is already underscored in the growing conflict over how to manage the most commonplace activities of our day-to-day life. Consider the fields of health care, architecture, biological research, energy development, agriculture, and the workplace as interesting indications of the new time polarization and time politics.

In medical schools, the standard approach to health care is still based on the idea of gaining control over the immediate environment in which a problem exists. Oftentimes, the body is treated like a machine that needs repair. In the process medical care is often cold, mechanical, and unfeeling. Doctors become more and more detached from patients, to the point where little if any physical contact exists between the two. In place of the warm, caring touch of another human being, the patient is monitored, scanned, probed, and screened by an impersonal array of sophisticated machinery. Modern medicine puts a premium on speed and efficiency. Surgical and pharmaceutical intervention is generally designed to quicken the pace of recovery. The goal is to supersede rather than complement the natural restorative rhythms.

Yet, there are signs that a new empathetic approach to medicine is emerging as a viable alternative. The holistic health movement emphasizes a nurturing, participating interaction between practitioner and patient. Patients are encouraged to identify with their own bodily rhythms and to work with, instead of against, the body's own restorative timetable. The body is not treated in isolation but rather as an integral part of the larger environment with which it is in constant rhythmic participation. The emphasis is on letting the entire environment assist in helping to restore the proper temporal relationships of the body.

An empathetic approach to medical treatment places greater attention on prevention over cure. For example, consider the two major diseases that confront industrial society today: cancer and heart disease. While these diseases have always existed, they have assumed epidemic proportions in the past several decades. A host of clinical studies point to a causal relationship between the poor nutritional, high-stress, carcinogenic environment we have created and the triggering of cancer and heart disease. The empathetic school of medicine focuses its research on finding ways to eliminate the source of the problem. Concern is directed toward reducing stress by slowing down the frenetic pace of life, changing nutritional habits, and cleaning up the polluted environment.

This holistic approach is in stark contrast to the more traditional forms of medicine that leave the source of the problems unattended, concentrating instead on surgical intervention and, in the future, genetic manipulation in order that people may continue to live in the polluted, high-stress environment that helps trigger the diseases in the first place.

202 · THE DEMOCRATIZATION OF TIME

Architecture provides still another example of the two different temporal orientations at work. Many contemporary architects dream of building a skyscraper, a monument of invincibility that can stand self-contained and isolated, jutting above its surroundings in princely relief. They are always in search of new building materials, new tools, and principles of organization that will allow them to build a fortress that can withstand and survive all possible assaults from the environment. The architects of skyscrapers pursue the idea of maximizing power and control over nature. Their works of art are designed to imperialize their environment, to expropriate their surroundings. Consider for a moment the fact that the Sears Tower in Chicago, the world's tallest building, functions at such an accelerated pace that it uses up more energy resources in a given twenty-four-hour day than the entire city of Rockford, Illinois, with its 147,000 inhabitants.

Then there are the new architects, whose approach to architecture is guided by very different considerations. They dream of building a passive solar home that is so elegant and unobtrusive, so meshed with the sequences, durations, and tempos of the natural environment, that it is difficult to distinguish where their time orientation leaves off and where nature's own timetables begin. Their interest is in developing materials, tools, and principles of design that are compatible with environmental rhythms. They see their building not as a fortress but as an environment within an environment, an extension of their surroundings that fully participates with the beats and periodicities of a larger setting. Their buildings participate with the heat and light of the sun, the cycles of the seasons, and the currents and tides of the winds and water. Their buildings belong to the environment. These architects pursue the idea of empathizing and participating with the rhythms of the natural world.

These two types of architects express two very different views of security in their pursuit of knowledge. For the designer of skyscrapers, security comes from building an impenetrable turret, a structure that can control and dominate its surroundings.

For the designer of the passive solar home, security comes from building a structure that can become part of a larger, already well-established community—the ecosystem.

In every branch of learning, a new generation of students and teachers is challenging the orthodox notion of power knowledge with the radically different idea of empathetic knowledge. For example, in university biology departments, genetic engineers are asking the "how" of nature so that they can design more efficient, more commerciably viable forms of life. On the other hand, there are the ecologists who are seeking to identify the subtle relationships and interactions between all living things so they can learn how to integrate the social and economic rhythms of society with the biological rhythms of the rest of the living kingdom.

In engineering departments, there are researchers working on elaborate nuclear fusion reactors, attempting to harness a storehouse of energy more powerful than the sun to give us even greater control over the environment than we have now. Then, there are the other breed of engineers who are working with wind, solar, and water power in an effort to develop appropriate technologies that are congenial to the rhythms of nature.

Agriculture provides yet another example of how time polarization is affecting the kinds of choices being made over how to manage the routine activities of society. High-technology farmers are constantly engaged in the pursuit of new, more exotic forms of plant and soil manipulation in order to exercise greater control over the forces of nature. For them, agriculture has become a battleground for introducing more sophisticated weapons, in the form of chemical fertilizers and pesticides, in an effort to exact as great a tribute as possible from the soil and plants under cultivation. In high-technology agriculture, the emphasis is on maximizing crop yield in the minimum amount of time. By infusing an array of laboratory-produced chemicals into the soil and by manipulating the genetic makeup of the plants to create more homogenous strains, it is possible to increase yield dramatically in the short run, but at the expense of the erosion and depletion of the soil base and a severe loss of genetic diversity in the plants in the long run.

In contrast, organic farmers are developing a sophisticated store of knowledge about the delicate balance of relationships that govern the plant cycles. They are introducing organic fertilizers and natural pest controls and are paying close attention to restoring the natural rhythms of production and recycling. Organic farmers see their role as nurturing rather than marshaling. Their concern is to preserve the soil base and the natural plant strains to ensure adequate reserves for future generations. In the short run, organic farmers are likely to get less yield than high-technology farmers. In the long run, their overall yield will be much higher because they allowed the soil and plants to replenish themselves over many cycles and seasons.

Nowhere is the contrast between the power approach and the empathetic approach to time orientation more vividly portrayed than in the workplace. The power time orientation emphasizes speed and predictability. Toward that end, personal involvement in the work cycle is reduced to a minimum. Individual workers are isolated from both the process and each other, forced to work within a narrow temporal band where they are denied access to past and future. Although they are caught up in the work cycle, they do not participate in it. Instead, they are swept up into the temporality imposed on them by machines and management, forced to conform to a rigidly defined set of sequences, durations, and rhythms over which they have little or no control.

A new generation of worker-owned and -operated companies is attempting to establish a radically different temporal orientation, one that integrates the total time needs of the individual workers with the time imperatives of the production process. In democratically run enterprises, each worker has a voice and a vote in the decisions that affect his or her life on the job. Group participation in every aspect of the work process ensures that each individual will have some say in establishing the tempo of work-related activities. Worker-run companies place greater emphasis on face-to-face interaction and on empathizing with the unique temporal needs of each employee. Many worker-run companies use appropriate small-scale technologies, tools that are designed to work in tandem with the natural biological rhythms of the human body.

Resource and waste management in worker-run companies is often more attuned to the timetable of the larger environment than in more traditional workplaces. Because the owner-employees also live in the communities in which they work, they are more likely to conserve the natural resources they depend on for their livelihood, as well as recycle wastes so as not to pollute their own communities. The rhythms of the workplace are made to coexist with the rhythms of the larger ecosystem and the extended life of the community. The cardinal rule in many such companies is not to produce, consume, and dispose at a pace that exceeds nature's own ability to replenish and recycle.

While most of the political life of the planet continues to revolve around the question of access to power, nascent constituencies and interest groups are beginning to make their presence felt in a revolutionary way. For them, access to power is less important than being part of a community. Their politics are the politics of empathy and participation. These new political constituencies are less interested in a voice inside the dominant political domicile and more interested in framing an entirely new social context outside the existing order.

The order they espouse is nonexclusionary. There are no time walls between people and no temporal fences separating humanity from the rest of nature. In this new political setting, consciousness gives rise to perspective and perspective gives rise to empathy, participation, and acceptance. In the old political framework, productivity, utility, and efficiency reign supreme. In the new political framework, the sacredness of all of life reigns supreme.

Many new movements have emerged in recent years, each embracing aspects of the empathetic time vision. The environmental movement, the animal-rights movement, the Judeo-Christian stewardship movement, the eco-feminist movement, the holistic health movement, the alternative agriculture movement, the appropriate technology movement, the bio-regionalism movement, the self-sufficiency movement, the economic democracy movement, the alternative education movement, and the disarmament movement come readily to mind.

If the activities of these new movements often appear apolitical, it is only because their methods of political engagement are often so different from what we have come to expect from new groups who have traditionally sought access to power. While there is some remaining residue of interest in traditional politics, in gaining recognition by and acceptance into the traditional political fold, there is also a marked turning away as these new groups set their sights on the politics of shared community, as opposed to the process of power politics.

We tend to think of political change in terms of reform or revolution. These new groups tend to think of political change more in terms of personal and collective transformation. Their interest is not in changing history but in redirecting consciousness. They believe that reorienting the psyche precedes reshaping the institutional environment. Many of the members of these newly emerging constituencies express their politics by the way they choose to live. For them, politics extends far beyond ballots and bills, referendums and recalls. Politics is a lifelong commitment to maintaining an empathetic context for life to flourish in.

All of the new empathetic movements share an abiding respect and appreciation for the rhythms of the natural world and are committed to the establishment of a social time order that is compatible with and complementary to the natural time order. For every one of these groups, the sacred rhythms of life come before the efficient rhythms of production. In this sense, their very presence poses a fundamental threat to political power, in both East and West.

Interestingly, there are signs that the new empathetic time politics is finding favor with a broader public. A Harris Poll conducted several years back found that by a margin of 79 percent to 17 percent, the public would place greater emphasis on teaching people how to live more with basic essentials "than on reaching a higher standard of living"; by 76 percent to 17 percent, a majority would opt for learning to get "pleasures out of nonmaterial experiences" rather than "satisfying our needs for more goods and services"; 59 percent believed we should be "putting a real effort into avoiding those things that cause pollution" over "finding ways to clean up the environment as the economy expands"; 82 percent would prefer to "improve those modes of travel we already have," while only 11 percent believed we should "develop ways to get more places faster"; and by 77 percent to 17 percent, Americans would prefer "spending more time getting to know each other better as human beings on a person-to-person basis" instead of "improving and speeding up our ability to communicate with each other through better technology." Finally, nearly two-thirds of the public said that "finding more inner and personal rewards from the work they do" is more important than "increasing the productivity of the work force"; "breaking up big things and getting back to more humanized living" should take precedence over "developing bigger and more efficient ways of doing things"; and "learning to appreciate human values more than material values" should be placed above "finding ways to create more jobs for producing more goods."[2]

Translating sentiments into commitments is never an easy affair, especially when it comes to the question of transforming the time consciousness of the culture. In a world caught up in speed, efficiency, and utility, how does one even begin to introduce the notion of empathy, participation, and communion into the body politic?

Making the transition from a politics based on power to a politics based on empathy requires a rethinking of the nature of free will. "To will" is to envision a future and to concentrate one's

energies toward the fulfillment of that vision. For much of Western history we have willed autonomy. We have envisioned a series
of fortress futures in which artificially controlled time frames are
imposed on the rhythms of nature. Our highest aim has been to
free ourselves from the fixed bonds of nature. We have long
associated free will with the loosening of our temporal ties with
the ecology of the planet and the strengthening of our bonds of
social autonomy. So much philosophical literature is taken up
with the intimate relationship between free will and autonomy
that for all practical purposes, they have come to mean virtually
the same thing. Exercising free will has come to mean exercising
autonomy. By defining free will in such a way, we have cast a
narrow vision over our political futures.

Now a new generation is beginning to inch its way toward a
new vision of the future. The politics of a postnuclear age replaces
the "will to power" with the "will to empathy," a transformational
politics for which there are only a few isolated examples in history
to look to as models.

Two thousand years ago, a carpenter from Galilee spread the
message of transformation in village settlements along the eastern frontiers of the Roman Empire. In this century, a black preacher
from the southern United States and an Indian barrister turned
mahatma inspired the empathetic vision of millions as they spread
the gospel of shared communion and called upon their country-
men to will a different kind of future into existence.

The empathetic approach to free will starts with exactly the
opposite assumptions of the power approach. Free will is no longer
measured by the degree of autonomy one exercises but rather by
the degree of communal participation and sharing that one experiences.

There is no reason why freedom need be any longer bound
to the notion of independence. One can be free to choose relationship over self-containment. The empathetic mind does just
that. It wills a future of stewardship over a future of manipulation. Belonging becomes more important than possessing.

The empathetic time horizon is saturated with responsibility and each act of will along the time line is dedicated to preserving the sacred rhythms of life. Nature, in the empathetic schema, is a life force, not a mere resource. Human beings choose to return to the temporal pace of nature's gait, abandoning their long quest to superimpose a set of artificial time constraints onto the biological world. Organic time ceases to be an obstacle to over-come and becomes, rather, a model to emulate as the political order attempts to reentrain the social clocks to the time cycles of nature.

In an empathetic time world, the mind places less emphasis on manipulative knowledge and more emphasis on revelatory knowledge. Manipulative knowledge gives us control but at the expense of wisdom. We become skilled craftsmen learning how to reshape surfaces without gaining any deep understanding of interiors. Manipulative knowledge is always exercised at the outer margins of reality. Revelatory knowledge is always experienced in the depths. In a world where the will is tutored only in the art of asking how to possess things, the future becomes an ever more alienating realm where technique substitutes for revelation. In such a willed world we become satisfied with explanations of how things can be exploited rather than why things exist. We become the technicians of the universe.

What, then, is our ultimate goal? Do we want the world to be remade, or revealed? The fact of the matter is, manipulative knowledge does not easily lend itself to revelatory experience. The former requires the cognition of control. The latter requires the cognition of surrender.

Revelation is experienced by a giving over, a reaching out. The essential why of things becomes revealed to us when we choose to surrender to them, to meet them on their own terms, to accept them for what they are. It is impossible to experience the essence of things when preoccupied with the thought of making them conform to our expectations. In a strange, ironic sense, the power time world we have created is steeped in illusion. We never meet things as they are. Instead we always experience reality in terms of how we would like it to be. In a power time world, essence gives way to expedience. We continually remake the world into our own image, and in the process, we never allow the time to experience the world unadorned by our illusions.

In an empathetic time world, our reality conforms to nature's. We trade partisanship for partnership and glory in our shared communion with the many time worlds that fill the universe.

We are indeed free to will two very different futures. In the first we seek to control the forces of nature and the lives of each other. In the second we seek to integrate ourselves back into the temporal bonds of the larger communities of life that make up the biosphere of the planet.

To will autonomy or to will community. To exercise power or to experience empathy. To control an artificial temporality or to rejoin the rhythmic world that is imprinted deep into the soul of our biological being. In the final analysis, we must ask what we want out of our future. Do we seek rulership or revelation? Mastery or self-discovery? Two futures await us, each accompanied by its own temporal mandate. The will to power, the will to empathy. The choice is ours. The time is now.

NOTES

PART I / THE TEMPORAL CONTEXT

1. THE NEW NANOSECOND CULTURE

1. Vincent E. Giuliano, "The Mechanization of Office Work," in *The Information Technology Revolution,* ed. Tom Forester (Cambridge: MIT Press, 1985), p. 301. Some experts have estimated that by 1988 nearly 60 percent of the American work force will be linked to electronic work stations. Presently there are more than 350,000 computers in the nation's schools, and at any given moment on Wall Street there are five times as many computers communicating with each other as people. See Craig Brod, *Technostress* (Reading, Mass.: Addison-Wesley, 1984), pp. 1–5.

2. Tracy Kidder, *The Soul of a New Machine* (Boston: Little, Brown, 1981), p. 137.

3. Sherry Turkle, *The Second Self: Computers and the Human Spirit* (New York: Simon & Schuster, 1984), p. 84.

4. Ibid.

5. Ibid., pp. 84–85.

6. Ibid., p. 87. Adds Turkle, "Everyone knows that the game is going to end 'sometime,' but 'sometime' is potentially infinite."

7. Brod, *Technostress*, pp. 115–16.

8. Geoff Simons, *Silicon Shock: The Menace of the Computer* (Oxford: Basil Blackwell, 1985), p. 165. For a more in-depth historical analysis of the psychological and social changes brought about by the "technologizing of the word," see Walter J. Ong, *Interfaces of the Word: Studies of the Evolution of Consciousness and Culture* (Ithaca, N.Y.: Cornell University Press, 1977); *Orality and Literacy* (New York: Methuen,

1982); Marshall McLuhan, *The Gutenberg Galaxy: The Making of Typographic Man* (Toronto: The University of Toronto Press, 1962); Harold A. Innis, *Empire and Communications* (Toronto: The University of Toronto Press, 1972).

9. Brod, *Technostress*, p. 94.

10. Ibid., p. 93. According to Brod, the technocentered computer compulsive loses all track of time—hours and minutes become irrelevant "as the task or program at hand *consumes* consciousness" (my emphasis).

11. Quoted in ibid.

12. Ibid., p. 105.

13. Ibid. Cf. Simons, p. 116, and Brod, p. 45.

14. David Bolter, *Turing's Man* (Chapel Hill: The University of North Carolina Press, 1984), p. 101.

15. Ibid.

16. Simons, *Silicon Shock,* p. 113.

17. Brod, *Technostress,* p. 26. A separate poll of 1,263 office workers in 1982 (sponsored by Verbatim Corporation, a manufacturer of floppy disks) revealed that 63 percent of those surveyed were concerned about eyestrain and 36 percent were concerned about backstrain. In addition, nearly eight in ten respondents called for better lighting conditions and 79 percent called for periodic rest breaks (p. 31).

18. Jerry Mander, "Six Grave Doubts About Computers," *Whole Earth Review* 44 (1985), pp. 14–15. See also David Burnham, *The Rise of the Computer State* (New York: Random House, 1982); Robert Sardello, "The Technological Threat to Education," *Teachers College Record* 85 (Summer 1984), pp. 631–39.

19. Quoted in Brod, *Technostress,* p. 126.

20. Ibid., p. 127. Jacques Ellul, in his seminal work *The Technological Society,* has pointed out that technique, by modifying and controlling both time and motion, radically disrupts and transforms human and biological time. See Ellul, *The Technological Society* (New York: Vintage Books, 1964), pp. 328–32.

21. John Davy, "Mindstorms in the Lamplight," *Teachers College Record* 85 (Summer 1984), p. 550. For a more general discussion on the quickened and harried pace of the modern child, see David Elkind, *The Hurried Child* (Reading, Mass.: Addison-Wesley, 1981).

22. Harriet K. Cuffaro, "Microcomputers in Education: Is Earlier Better?" *Teachers College Record* 85 (Summer 1984), p. 561.

23. Ibid.

24. Ibid., p. 562. See also Hubert L. Dreyfus and Stuart E. Dreyfus,

"Putting Computers in Their Proper Place: Analysis Versus Intuition in the Classroom," *Teachers College Record* 85 (Summer 1984), pp. 578–601; Hope Jensen Leichter, "A Note on Time and Education," *Teachers College Record* 81 (Spring 1980), pp. 360–70. Cf. Seymour Papert, *Mindstorms: Children, Computers, and Powerful Ideas* (New York: Basic Books, 1980).

25. Quoted in Brod, *Technostress,* pp. 129–30.
26. Quoted in ibid.
27. Mander, "Six Grave Doubts," pp. 18–19.

2. CHRONOBIOLOGY: THE CLOCKS THAT MAKE US RUN

1. G. J. Whitrow, *The Natural Philosophy of Time* (Oxford: Oxford University Press, 1980), p. vii. Some three decades later Karl von Frisch would add, "I know of no other living creature that learns so easily as the bee when, according to its 'eternal clock,' to come to the table." Von Frisch, *The Dance Language and Orientation of Bees,* trans. L. E. Chadwick (Cambridge: Harvard University Press, 1967), p. 253.

2. Colin Pittendrigh, "On Temporal Organization in Living Systems," in *The Future of Time,* eds. Henry Yaker, Humphrey Osmond, and Frances Cheek (New York: Anchor/Doubleday, 1972), pp. 196–97.

3. Ibid., p. 197.

4. This pioneering study was completed in 1955 and reported in the German science journal *Naturwissenschaften* by biologist M. Renner. Renner, *Naturwissenschaften* 42 (1955), pp. 540–41. See also Whitrow, *The Natural Philosophy of Time,* p. 133.

5. There is much well-known and well-observed data on birds returning each year to their summer haunts on exactly the same date. Besides the swallows of Capistrano, the less-celebrated Ohio turkey vulture returns each spring from the south on precisely the fifteenth of March.

6. Whitrow, *The Natural Philosophy of Time,* p. 141. In 1758 another Frenchman, Henri-Louis Duhamel du Monceau, repeated Mairan's experiment and confirmed not only the plant's ability to tell time in darkness but also the plant's ability to do this independent of temperature.

7. Ibid., pp. 143–44.

8. C. P. Richter, "Psychopathology of Periodic Behavior in Animals and Man," in *Comparative Psychopathology,* eds. J. Zubin and H. F. Hunt (New York: Grune & Stratton, 1967). See also Gay Gaer Luce, *Biological Rhythms in Human and Animal Physiology* (New York: Dover Publications, 1971), p. 15; and *Body Time: Physiological Rhythms and*

214 * Notes

Social Stress (New York: Bantam, 1973), pp. 138, 190–91, 213–15.

9. John Orme, "Time: Psychological Aspects—Time, Rhythms, and Behavior," in *Making Sense of Time,* eds. Tommy Carlstein, Don Parkes, and Nigel Thrift (New York: John Wiley & Sons, 1978), p. 67.

10. Leonard W. Doob, *Patterning of Time* (New Haven: Yale University Press, 1971), pp. 69–70. See also Jurgen Aschoff, "Circadian Rhythms in Man," *Science* 148 (1965), pp. 1427–32. On a more social or psychological level, the idea of doing the "right thing at the right time" corresponds with the Greek concept of *kairos.* Kairotic time is both participatory and empathetic.

11. Whitrow, *The Natural Philosophy of Time,* p. 146.

12. Ibid.

13. Ibid. There is a whole range of periodic phenomena for which there is no ordinarily observable external cue or "time-giver" (*Zeitgeber*). These endogenous biological rhythms may not be determined solely by the "environmental cycle," but by the animal's own continuously running clock. See, for example, Martin C. Moore-Ede, et al., *The Clocks That Time Us: Physiology of the Circadian Timing System* (Cambridge: Harvard University Press, 1982), pp. 51–81.

14. E. T. Pengelley and K. C. Fisher, "The Effect of Temperature and Photoperiod on the Yearly Hibernating Behavior of Captive Golden-Mantled Ground Squirrels," *Canadian Journal of Zoology* 41 (1963), pp. 1103–20.

15. See F. A. Brown, "A Unified Theory for Biological Rhythms," in *Circadian Clocks* (Amsterdam: North Holland, 1965), pp. 231–61. An entire chapter is devoted to Brown's pioneering work in Richie Ward, *The Living Clocks* (New York: Alfred A. Knopf, 1971), pp. 259–78.

16. Cited in Luce, *Biological Rhythms in Human and Animal Physiology,* pp. 12–13.

17. Orme, "Time," p. 66.

18. Martin C. Moore-Ede, Frank M. Sulzman, and Charles A. Fuller, *The Clocks That Time Us* (Cambridge: Harvard University Press, 1982), pp. 319–20.

19. The social and psychological impact of daylight savings time has been far greater than one might imagine. For a detailed study of the social and historical origins of daylight savings time, see Oliver B. Pollak, "Efficiency, Preparedness, and Conservation: The Daylight Savings Time Movement," *History Today* 31 (1981), pp. 5–9.

20. See Len Hilts, "Clocks That Make Us Run," *Omni* (September 1984), pp. 52–54. See also W. Herbert, "Punching the Biological Time-clock," *Science News,* 31 July 1982, p. 69.

21. Hilts, "Clocks," p. 52.

22. J. N. Mills, "Transmission Processes Between Clock and Manifestations," in *Biological Aspects of Circadian Rhythms,* ed. J. N. Mills (New York: Plenum Press, 1973), p. 57.

23. Luce, *Biological Rhythms in Human and Animal Physiology,* p. 48. This particular rhythm (the glycogen curve) is very important for physicians in the treatment of diabetics and in understanding responses to poisons and drugs.

24. Ibid., p. 10. See also Luce, *Body Time,* pp. 15–16, 31–32.

25. Ibid., p. 56.

26. Ibid. See also R. T. Wilkinson, "Evoked Response and Reaction Time," *Acta Psychologia* 27 (1967), pp. 235–45.

27. Hudson Hoagland, "Some Biochemical Considerations of Time," in *The Voices of Time,* ed. J. T. Fraser (Amherst: The University of Massachusetts Press, 1981), pp. 312–29.

28. John Cohen, *Psychological Time in Health and Disease* (Springfield, Ill.: Charles C. Thomas, 1967), p. 30. See also Michel Siffre, *Beyond Time* (New York: McGraw-Hill, 1964).

29. Luce, *Biological Rhythms in Human and Animal Physiology,* p. 5.

30. Quoted in Hilts, "Clocks," p. 100.

31. Quoted in ibid., p. 52.

32. Ibid.

33. J. G. Bohlen, "Circadian and Cirannual Rhythms in Eskimos" (Prospectus of Research, University of Wisconsin, Madison, 1969).

34. Joost A. M. Meerloo, *Along the Fourth Dimension* (New York: John Day Co., 1970), p. 79.

35. Boyce Rensberger, "Spring Fever Not Imaginary, Study Finds," *Washington Post,* 11 March 1985, p. A5.

36. Ibid.

37. Meerloo, *Along the Fourth Dimension,* p. 67. See also K. Hammer, "Experimental Evidence for the Biological Clock," in *The Voices of Time,* ed. Fraser; Sharon Sharp, "Biological Rhythms and the Timing of Death," *Omega* 12 (1981–1982), pp. 15–23; George L. Engel, "The Need for a New Model: A Challenge for Biomedicine," *Science* 196 (8 April 1977), pp. 129–36.

38. W. S. Condon and W. D. Ogston, "Sound Film Analysis of Normal and Pathological Behavior Patterns," *The Journal of Nervous and Mental Disease* 143 (1966), pp. 338–46.

39. W. S. Condon, "A Primary Phase in the Organization of Infant Responding Behavior," in *Studies in Mother-Infant Interaction,* ed.

216 ·   *Notes*

H. R. Schaffer (New York: Academic Press, 1977), p. 167. See also Alexander Thomas, Stella Chess, and Herbert G. Birch, "The Origin of Personality," *Scientific American* 223 (1970), p. 104; Luce, *Biological Rhythms in Human and Animal Physiology,* p. 38; Esther Thelen, "Rhythmical Behavior in Infancy: An Ethological Perspective," *Developmental Psychology* 17 (1981), pp. 237–57.

40. See, for example, Sigmund Freud, *Civilization and Its Discontents* (London: Hogarth Press, 1930). Cf. N. O. Brown, *Life Against Death,* 2nd ed. (Middletown, Conn.: Wesleyan University Press, 1986); Herbert Marcuse, *Eros and Civilization* (New York: Vintage Books, 1962). There are a number of works dealing specifically with the psychology of time consciousness. Some of the better ones are: Paul Fraisse, *The Psychology of Time,* trans. Jennifer Leith (New York: Harper & Row, 1963); Robert E. Ornstein, *On the Experience of Time* (Baltimore: Penguin Books, 1969); Bernard S. Gorman and Alden E. Wessman, eds., *The Personal Experience of Time* (New York: Plenum Press, 1977); Frederick Towne Melges, *Time and the Inner Future* (New York: John Wiley & Sons, 1982); Matthew Edlund, *Psychological Time and Mental Illness* (New York: Gardner Press, 1986).

41. Edmund Bergler and Géza Roheim, "Psychology of Time Perception," *Psychoanalytic Quarterly* 15 (1946), p. 198. See J. T. Fraser's comments on Freud in Fraser, *Of Time, Passion, and Knowledge* (New York: George Braziller, 1975), pp. 287–94. See also Edmund Bergler, "Psychoanalysis of the Ability to Wait and of Impatience," *Psychoanalytic Review* 26 (1939), pp. 11–32; Louis Scheider, "The Deferred Gratification Pattern: A Preliminary Study," *American Sociological Review* 18 (1953), pp. 142–49; M. Guy Thompson, *The Death of Desire* (New York: New York University Press, 1985), pp. 118–35.

42. Freud, *Civilization and Its Discontents,* p. 63. For a more existential look at the relationship between the libidinous urge and temporality, see Alphonso Lingis, *Libido: The French Existential Theories* (Bloomington: The University of Indiana Press, 1985), pp. 66–69, 75–80.

43. Lawrence Joseph Stone and Joseph Church, *Childhood and Adolescence: A Psychology of the Growing Person* (New York: Random House, 1957), p. 184. See also Thomas J. Cottle and Stephen Klineberg, *The Present of Things Future* (New York: The Free Press/Macmillan, 1974), pp. 70–101.

44. See Melvin Wallace and Albert I. Rabin, "Temporal Experience," *Psychological Bulletin* 57 (1960), pp. 213–36.

45. Cottle and Klineberg, *The Present of Things Future,* p. 93.

3. ANTHROPOLOGICAL TIME ZONES

1. Daniel J. Boorstin, *The Discoverers* (New York: Random House, 1983), p. xvii. John O'Neill, in his sensitive critique of current developments in critical theory, confronts the notion of human time *within* the historical context. We must remember and flesh out the past if we are to discuss the present or future critically. The role of remembrance is a crucial tool of the socially conscious historian, as O'Neill points out; it should not be used carelessly or without the aid of an acute *historical* memory. Milan Kundera, in the literary genre, has also addressed the "politics of memory." See John O'Neill, "Critique and Remembrance," in *On Critical Theory*, ed. John O'Neill (New York: Seabury Press, 1976), pp. 1–11; Milan Kundera, *The Book of Laughter and Forgetting* (New York: Penguin Books, 1980). For a discussion written specifically about temporality and memory, see Steinar Kvale, "The Temporality of Memory," *The Journal of Phenomenological Psychology* 5 (1974), pp. 7–31.

2. There is an interesting literature dealing with time, temporality, and the increasing "punctuality of the family." The ecology of the family, parent to child, sibling to sibling, has many fascinating temporal dimensions. See, for example, Fred Darnley, "Periodicity in the Family: What Is It and How Does It Work?" *Family Relations* 30 (1981), pp. 31–37; Ronald Cromwell, Bradford Keeney, and Bert N. Adams, "Temporal Patterning in the Family," *Family Process* 15 (1975), pp. 345–47; David Kantor and William Lehr, *Inside the Family: Toward a Theory of Family Process* (San Francisco: Jossey-Bass Publishers, 1975), especially pp. 43–84.

3. Edward T. Hall, *The Dance of Life: The Other Dimension of Time* (New York: Anchor/Doubleday, 1983), p. 170.

4. Lois Pratt, "Business Temporal Norms and Bereavement Behavior," *American Sociological Review* 46 (1981), pp. 317–33.

5. Robert McCaffery, *Managing the Employee Benefits Program* (New York: American Management Association, 1972), p. 125.

6. Pratt, "Business Temporal Norms," p. 326.

7. Ibid., pp. 330–31.

8. Ibid, p. 327.

9. Ibid.; see also Robert Fulton, "The Sacred and the Secular: Attitudes of the American Public Toward Death and Funeral Directors," in *Death and Identity*, ed. Robert Fulton (New York: John Wiley & Sons, 1965), pp. 89–105; David Stannard, *The Puritan Way of Death* (New York: Oxford University Press, 1977).

10. Quoted in Thomas Cottle and Stephen Klineberg, *The Present of*

Things Future (New York: The Free Press/Macmillan, 1974), p. 168.

11. P. M. Bell, "Sense of Time," *New Science* 15 (1975), p. 406. Similar behavior has been exhibited by Australian aboriginal children. Even though these children on the whole have a similar mental capacity to white children, they find it extremely difficult to tell time by the clock. See, for example, C. J. Whitrow, *The Natural Philosophy of Time* (Oxford: Oxford University Press, 1980), p. 55.

12. Pitirim A. Sorokin and Robert K. Merton, "Social Time: A Methodological and Functional Analysis," *The American Journal of Sociology* 42 (1937), p. 619.

13. Quoted in E. P. Thompson, "Time, Work-Discipline, and Industrial Capitalism," *Past and Present* 38 (1967), p. 58. Any in-depth study of primitive culture should illuminate the primitive's profound relationship to time and natural events. Mircea Eliade, for example, has widely discussed this realm in his philosophical anthropology via "sacred" and "profane" temporal categories. Stanley Diamond's anthropological "search for the primitive" has also yielded some insight into the time consciousness of early man. See Eliade, *The Sacred and the Profane,* trans. Willard Trask (New York: Harcourt, Brace & World, 1959), pp. 68–116; and Stanley Diamond, *In Search of the Primitive* (New Brunswick, N.J.: Transaction Books, 1981), pp. 203–26.

14. Quoted in Amos Hawley, *Human Ecology* (New York: Ronald Press, 1950), pp. 296–97.

15. Robert Levine with Ellen Wolff, "Social Time: The Heartbeat of Culture," *Psychology Today* 19 (1985), pp. 28–30.

16. Ibid., p. 30.

17. Eviatar Zerubavel, *Hidden Rhythms: Schedules and Calendars in Social Life* (Chicago: The University of Chicago Press, 1981), pp. 49–50.

18. A. Irving Hallowell, "Temporal Orientation in Western Civilization and in a Preliterate Society," *American Anthropologist* 39 (1937), p. 652.

19. Hawley, *Human Ecology*, p. 300.

20. Quoted in Sorokin and Merton, "Social Time," pp. 621–22. The Taoist vision is a highly ecological one in that the rhythms of the earth are carefully observed and respected. For a temporal, historical, and religious exegesis on Taoist cosmology, see R. G. H. Siv, *Chi: A Neo-Taoist Approach to Life* (Cambridge: MIT Press, 1974). For the relationship between ecology and Taoism, see Russell Goodman, "Taoism and Ecology," *Environmental Ethics* 2 (1980), pp. 73–80.

21. Sorokin and Merton, "Social Time," pp. 621–22.

22. Hall, *The Dance of Life,* p. 172.
23. Ibid.
24. Ibid.
25. Quoted in Zerubavel, *Hidden Rhythms,* p. 16.
26. Robert Knapp and John Garbutt, "Time Imagery and the Achievement Motive," *Journal of Personality* (1958), pp. 427–28. See also Knapp and Garbutt, "Variation in Time Descriptions and Need Achievement," *The Journal of Social Psychology* 67 (1965), pp. 269–72.
27. Hallowell, "Temporal Orientation," pp. 649–50.
28. Ibid.
29. Quoted in Levine with Wolff, "Social Time," p. 34.
30. Ibid.
31. Hall, *The Dance of Life,* p. 154.
32. Ibid., pp. 154–55.
33. Ibid., p. 155.
34. Ibid., p. 150.
35. Ibid.
36. Oren Lyons, "An Iroquois Perspective," in *American Indian Environments: Ecological Issues in Native American History,* eds. Christopher Yecsey and Robert W. Venables (New York: Syracuse University Press, 1980).

PART II / DIVIDING THE TIME PIE

4. CALENDARS AND CLOUT

1. Quoted in Stephen Kern, *The Culture of Time and Space 1880–1918* (Cambridge: Harvard University Press, 1983), pp. 19–20.
2. Pitirim A. Sorokin, *Sociocultural Causality, Space, Time* (New York: Russell and Russell, 1964), p. 173. See also Rudolf Rezsöházy, "The Concept of Social Time: Its Role and Development," *International Social Science Journal* 24 (1972), pp. 26–36.
3. Eviatar Zerubavel, *Hidden Rhythms: Schedules and Calendars in Social Life* (Chicago: The University of Chicago Press, 1981), p. 73.
4. Quoted in ibid., p. 74.
5. Ibid.
6. Ibid. The Hebrew Sabbath has served a most significant symbolic function throughout its history. Not only does the observance of the Sabbath reflect the fundamental need of the Jews to distinguish them-

selves from their Gentile environment, it also emphasizes their uniqueness as a social group.

7. Daniel J. Boorstin, *The Discoverers* (New York: Random House, 1983), pp. 597–98.

8. Lawrence Wright, *Clockwork Man* (New York: Horizon Press, 1969), p. 47.

9. Eviatar Zerubavel, "Easter and Passover: On Calendars and Group Identity," *American Sociological Review* 47 (April 1982), pp. 287–88.

10. Quoted in ibid., p. 287.

11. Ibid., p. 288. See also Eviatar Zerubavel's recent work on the social and political function of the seven-day week, *The Seven Day Circle* (New York: The Free Press/Macmillan, 1985).

12. Thomas Darby, *The Feast: Meditation on Politics and Time* (Toronto: The University of Toronto Press, 1982), p. 59.

13. Zerubavel, *Hidden Rhythms,* p. 85. For a more traditional, albeit enlightening, interpretation of the political secularization of the public-religious sphere, see Hannah Arendt, *Between Past and Future* (New York: Penguin Books, 1977), pp. 63–75.

14. Zerubavel, *Hidden Rhythms,* pp. 89–90.

15. Ibid., pp. 90–91.

16. Ibid., p. 92.

17. See, for example, Sebastian de Grazia, *Of Time, Work, and Leisure* (New York: The Twentieth Century Fund, 1962), p. 119.

5. SCHEDULES AND CLOCKS

1. George Woodcock, "The Tyranny of the Clock," in *Politics* 1 (1944), pp. 265–66.

2. Sebastian de Grazia, *Of Time, Work, and Leisure* (New York: The Twentieth Century Fund, 1962), p. 41.

3. Quoted in ibid., p. 54.

4. Justin McCann, trans., *The Rule of St. Benedict* (London: Sheed and Ward, 1970), chap. 48.

5. Ibid.

6. Eviatar Zerubavel, *Hidden Rhythms: Schedules and Calendars in Social Life* (Chicago: The University of Chicago Press, 1981), p. 33.

7. McCann, *The Rule of St. Benedict,* chap. 22.

8. Zerubavel, *Hidden Rhythms,* p. 32.

9. Reinhard Bendix, *Max Weber* (Garden City, N.Y.: Anchor, 1962), p. 318.

10. Quoted in Lawrence Wright, *Clockwork Man* (New York: Horizon Press, 1969), p. 208.

11. Daniel J. Boorstin, *The Discoverers* (New York: Random House, 1983), p. 38.

12. Wright, *Clockwork Man,* p. 62.

13. David Landes, *Revolution in Time* (Cambridge: Harvard University Press, 1983), p. 194.

14. F. C. Haber, "The Cathedral Clock and the Cosmological Metaphor," in *The Study of Time II,* eds. J. T. Fraser and N. Lawrence (New York: Springer-Verlag, 1975), p. 401.

15. Ibid.

16. Woodcock, "Tyranny of the Clock," p. 266. The science, history, and art of clock construction is commonly referred to as horology. In their study of the history and evolution of timepieces, horologists have amassed a wealth of relevant material. See, for example, Carlo Cipolla, *Clocks and Culture 1300–1700* (New York: Walker, 1967); Joseph Needham, Wang Ling, and Derek J. de Solla Price, *Heavenly Clockwork: The Great Astronomical Clocks of Medieval China* (Cambridge: At the University Press, 1960); Daniel W. Hering, *The Lure of the Clock* (New York: New York University Press, 1932); Eric Bruton, *The History of Clocks and Watches* (London: Orbis, 1979).

17. Wright, *Clockwork Man,* p. 55.

18. Ibid.

19. Lewis Mumford, *Technics and Civilization* (New York: Harcourt, Brace and World, 1934), p. 15.

20. Landes, *Revolution in Time,* p. 16.

21. Ibid., pp. 72–73.

22. Jacques Le Goff, *Time, Work, and Culture in the Middle Ages* (Chicago: The University of Chicago Press, 1980), p. 35.

23. Quoted in Jack Goody, "Time: Social Organization," in *International Encyclopedia of the Social Sciences,* vol. 16, ed. David Sills (New York: The Free Press/Macmillan, 1968), pp. 38–39.

24. See Mumford, *Technics and Civilization,* p. 16.

25. Le Goff, *Time, Work, and Culture in the Middle Ages,* pp. 49–50.

26. Landes, *Revolution in Time,* p. 12. For a much broader discussion on the role of the clock in shaping Western consciousness and culture, see Samuel Macey, *Clocks and the Cosmos* (Hamden, Conn.: Archon Books, 1980).

6. TIME SCHEDULES AND FACTORY DISCIPLINE

1. Sebastian de Grazia, *Of Time, Work, and Leisure* (New York: The Twentieth Century Fund, 1962), p. 59.

2. Sidney Pollard, *The Genesis of Modern Management* (Cambridge: Harvard University Press, 1965), p. 161.
3. Ibid., p. 162.
4. Ibid., p. 161.
5. Ibid., p. 173.
6. Daniel J. Boorstin, *The Discoverers* (New York: Random House, 1983), p. 72.
7. Wilbert E. Moore, *Man, Time, and Society* (New York: John Wiley & Sons, 1963), p. 28.
8. Lawrence Wright, *Clockwork Man* (New York: Horizon Press, 1969), pp. 118–19.
9. E. P. Thompson, "Time, Work-Discipline, and Industrial Capitalism," *Past and Present* 38 (1967), p. 81.
10. Ibid., p. 82.
11. Ibid.
12. Ibid., p. 82.
13. Pollard, *The Genesis of Modern Management*, p. 184.
14. Ibid., p. 182.
15. Anonymous, *Chapters in the Life of a Dundee Factory Boy* (Dundee, England, 1887), p. 10.
16. Pollard, *The Genesis of Modern Management*, p. 185. In the Cumberland mines, for example, children started work at the ages of 5 to 7, and as late as 1842, 250 of the 1,400 workers in the Lonsdale mines were under 18.
17. Ibid., p. 188.
18. Ibid.
19. Ibid., p. 191.
20. Thompson, "Time, Work-Discipline, and Industrial Capitalism," p. 87.
21. Ibid.
22. Quoted in ibid, p. 84.
23. Ibid., p. 85.
24. Wright, *Clockwork Man*, p. 121.

7. PROGRAMS AND COMPUTERS

1. See Gene Bylinsky and Alicia Hills Moore, "Flexible Manufacturing Systems," in *The Information Technology Revolution,* ed. Tom Forester (Cambridge: MIT Press, 1985), pp. 287–89. Although Japan's "fifth generation" is making remarkable "advances" in the computerization of the factory, their recent research into artificial intelligence should also

be noted. See, for example, Edward A. Feigenbaum and Pamela Mc-
Corduck, *The Fifth Generation: Artificial Intelligence and Japan's Computer Challenge to the World* (New York: New American Library, 1984).
2. Ibid.
3. Marshall McLuhan, *Understanding Media: The Extensions of Man* (New York: McGraw-Hill, 1965), p. 346. Note McLuhan's discussion on clocks in this same text, pp. 145–56.
4. David Bolter, *Turing's Man* (Chapel Hill: The University of North Carolina Press, 1984), p. 101.
5. Ibid., p. 102.
6. Ibid.
7. Ibid., pp. 102–103.
8. Sherry Turkle, *The Second Self: Computers and the Human Spirit* (New York: Simon & Schuster, 1984), p. 13.

8. THE EFFICIENT SOCIETY

1. Harry Braverman, *Labor and Monopoly Capital* (New York: Monthly Review Press, 1974), p. 70.
2. Adam Smith, *The Wealth of Nations* (New York: New Modern Library, 1937), p. 7.
3. Samuel L. Macey, *Clocks and the Cosmos: Time in Western Life and Thought* (Hamden, Conn.: Archon Books, 1980), p. 35.
4. Quoted in ibid.
5. Daniel Bell, "The Clock Watchers: Americans at Work," *Time*, 8 September 1975, p. 55.
6. Macey, *Clocks and the Cosmos*, p. 36.
7. Bell, "The Clock Watchers," p. 55.
8. Braverman, *Labor and Monopoly Capital*, p. 88.
9. Ibid., p. 67.
10. Ibid., p. 88.
11. Frederick Taylor, *The Principles of Scientific Management* (New York: W. W. Norton, 1947), pp. 37–38.
12. Ibid., pp. 235–36.
13. Ibid., pp. 39, 63.
14. Braverman, *Labor and Monopoly Capital*, pp. 173–74.
15. Ibid., p. 177.
16. Ibid., p. 178.
17. Ibid., p. 321.
18. Stephen Kern, *The Culture of Time and Space 1880–1918* (Cambridge: Harvard University Press, 1983), p. 12.

19. Ibid.
20. David S. Landes, *Revolution in Time* (Cambridge: Harvard University Press, 1983), pp. 285–86.
21. Ibid., p. 286.
22. Ibid.
23. Kern, *The Culture of Time and Space*, p. 13.
24. Robert H. Lauer, "Temporality and Social Case of the 19th Century in China and Japan," *The Sociological Quarterly* 14 (1973), pp. 452–53.
25. Ibid., p. 453. See also Chung-ying Cheng, "Greek and Chinese Views on Time and the Timeless," *Philosophy East and West* 24 (1974), pp. 155–59.
26. Ibid., p. 454. For an excellent survey of the Chinese time consciousness and China's strong temporal orientation to the past, see Joseph R. Levenson, *Confucian China and Its Modern Fate: A Trilogy* (Berkeley: The University of California Press, 1968).
27. Quoted in ibid.
28. Ibid.
29. Ibid., p. 457.
30. Nyozekan Hasegawa, *The Japanese Character,* trans. John Bestor (Tokyo: Kodansha International, 1965), pp. 101–102.
31. Lauer, "Temporality and Social Case," p. 461. Several Sinologists have pointed out that Japan's Tokugawa period is the Oriental equivalent to the West's "Protestant ethic" era. See, for example, Robert Bellah, *Tokugawa Religion* (New York: The Free Press/Macmillan, 1957). It appears, evidently, that nearly all societies that embrace industrialization develop the kinds of attitudes that Weber discusses in his classic work.
32. Ibid., p. 458. See also W. Caudhill and H. A. Scarr, "Japanese Value Orientations and Culture Change," *Ethology* 1 (1962), pp. 53–91; Marius Jansen, *Changing Japanese Attitudes Towards Modernization* (Princeton, N. J.: Princeton University Press, 1965); Hajime Nakamura, "Time in Indian and Japanese Thought," in *The Voices of Time,* ed. J. T. Fraser (Amherst: The University of Massachusetts Press, 1981), pp. 77–91.
33. Craig Brod, *Technostress* (Reading, Mass.: Addison-Wesley, 1984), pp. 39–40. For a general summary of the effects of automation with respect to time, see George Soule, *What Automation Does to Human Beings* (New York: Arno Press, 1977), pp. 86–101, 130–46.
34. Christopher Rawlence, ed., *About Time* (London: Jonathan Cape, 1985), p. 39.
35. Ibid., p. 39.

36. Brod, *Technostress*, p. 45.

37. Ibid.

38. Ibid.

39. Rawlence, *About Time*, p. 39.

40. Vic Sussman, "Going Nowhere Fast," *Washington Post Magazine*, 7 September 1986, p. 77.

PART III / THE POLITICS OF PARADISE

9. THE TIMELESS STATE

1. Frederick L. Polak, *The Image of the Future*, vol. 1 (Leyden, Netherlands: A. W. Sitjhoff, 1961), p. 49.

2. S. G. F. Brandon, *History, Time and Deity* (New York: Barnes & Noble, 1965), p. 206.

3. Jean Guitton, *Man in Time* (Notre Dame, Ind.: The University of Notre Dame Press, 1966), pp. 28–35. See also Martin Buber, *Paths in Utopia* (Boston: Beacon Press, 1958); Norman Cohn, *The Pursuit of the Millennium* (New York: Harper & Row, 1961); Wilbert E. Moore, *Man, Time, and Society* (New York: John Wiley & Sons, 1963), pp. 151–52; John Passmore, *The Perfectibility of Man* (New York: Charles Scribner's Sons, 1970), pp. 304–27.

4. Andre Neher, "The View of Time and History in Jewish Culture," in *Cultures and Time*, eds. L. Gardet et al. (Paris: The Unesco Press, 1976), pp. 149–55.

5. Polak, *The Image of the Future*, pp. 174–80.

6. Neher, "The View of Time," pp. 155–56.

7. Polak, *The Image of the Future*, p. 177.

8. Daniel J. Boorstin, *The Discoverers* (New York: Random House, 1983), p. 567.

9. Polak, *The Image of the Future*, p. 185.

10. Ibid., p. 159.

11. Ibid., p. 163.

12. Mircea Eliade, *The Myth of the Eternal Return* (Princeton, N.J.: Princeton University Press, 1954), pp. 105–107.

13. Ibid., p. 107. See also J. L. Russell, "Time in Christian Thought," in *The Voices of Time*, ed. J. T. Fraser (Amherst: The University of Massachusetts Press, 1981), pp. 59–76.

14. Polak, *The Image of the Future*, p. 152. See Rom. 14:17; 1 Cor. 15:50.

15. Brandon, *History, Time, and Deity*, pp. 26–27.

16. Ibid., p. 27.

17. Polak, *The Image of the Future*, pp. 168–71.

18. Ibid., pp. 168–70.

19. Ibid., pp. 147–51.

20. Wendell Bell and James A. Mau, "Images of the Future: Theory and Research Strategies," in *Theoretical Sociology: Perspectives and Developments*, ed. Edward A. Tiryakian (New York: Appleton, Century, Crofts Meredith, 1970), p. 213.

10. THE IMAGE OF PROGRESS

1. Frederick L. Polak, *The Image of the Future*, vol. 1 (Leyden, Netherlands: A. W. Sitjhoff, 1962), p. 202.

2. Jacques Le Goff, *Time, Work, and Culture in the Middle Ages* (Chicago: The University of Chicago Press, 1980), p. 59.

3. Ibid., pp. 59–61.

4. Ricardo J. Quinones, *The Renaissance Discovery of Time* (Cambridge: Harvard University Press, 1972), p. 5.

5. Le Goff, *Time, Work, and Culture*, p. 51.

6. Ibid., p. 29.

7. Ibid., p. 61. *Nihil inde sperantes.*

8. Ibid.

9. Ibid., p. 30.

10. Quoted in William J. Bouwsma, "Anxiety of the Formation of Early Modern Culture," in *After the Reformation*, ed. Barbara Coralament (Philadelphia: The University of Pennsylvania Press, 1980), p. 230.

11. Francis Bacon, *Novum Organum*, book 1, aphorism 2.

12. Jean Houston, "Prometheus Rebound: An Inquiry into Technological Growth and the Psychology of Change," in *Alternatives to Growth*, ed. Dennis Meadows (Cambridge: Bollingen, 1977), p. 274.

13. John Herman Randall, *The Making of the Modern Mind* (Cambridge: Houghton Mifflin, 1940), p. 241.

14. Quoted in ibid., pp. 241–42.

15. Quoted in Leo Strauss, *Natural Rights and History* (Chicago: The University of Chicago Press, 1953), p. 258.

16. Adam Smith, *An Inquiry into the Nature and Causes of the Wealth of Nations* (London: Methuen, 1961), p. 475.

17. Stephen Kern, *The Culture of Time and Space 1880–1918* (Cambridge: Harvard University Press, 1983), p. 95.

18. J. B. Priestley, *Man and Time* (New York: Dell Publishing Co., 1964), p. 162.

19. Lawrence Wright, *Clockwork Man* (New York: Horizon Press, 1969), p. 126.

20. Ibid., p. 128.

21. Ibid., p. 149.

22. Daniel Bell, "The Clock Watchers: Americans at Work," *Time,* 8 September 1975, p. 55.

23. Wright, *Clockwork Man,* p. 154.

24. Ibid., p. 159.

25. Ibid., p. 158.

26. Ibid., p. 112. According to Wright, London was said to be the "worst-lit" capital in Europe until 1736 when the city began to utilize fish-oil lamps. By the end of the decade, there "were 15,000 'parish lamps' burning fish oil throughout the year" (p. 107). "Even then," says Wright, "night life was not for the respectable citizen."

27. Quoted in Kern, ibid., p. 29.

28. Marquis de Condorcet, "Outline of an Historical View of Progress of the Human Mind," quoted in John Hallowell, *Main Currents in Modern Political Thought* (New York: Holt, Rinehart and Winston, 1950), p. 132.

11. THE VISION OF SIMULATED WORLDS

1. Sherry Turkle, *The Second Self: Computers and the Human Spirit* (New York: Simon & Schuster, 1984), p. 267.

2. Quoted in Craig Brod, *Technostress* (Reading, Mass.: Addison-Wesley, 1984), p. 8.

3. David Bolter, *Turing's Man* (Chapel Hill: The University of North Carolina Press, 1984), pp. 187–88.

4. Edward A. Feigenbaum, *The Fifth Generation: Artificial Intelligence and Japan's Computer Challenge to the World* (New York: New American Library, 1984), p. 251.

5. Ibid., p. 15. For a more philosophical treatment of artificial intelligence, see Herbert A. Simon, *The Sciences of the Artificial* (Cambridge: MIT Press, 1969); Gary Clark, "Artificial Intelligence and Philosophers," *Philosophy Today* 29 (1985), pp. 326–31. For three excellent discussions summarizing recent trends in A.I., see Frank Rose, *Into the Heart of Mind: An American Quest for Intelligence* (New York: Harper & Row, 1984); Roger C. Schank, *The Cognitive Computer: On Language, Learning, and Artificial Intelligence* (Reading, Mass.: Addison-Wesley, 1984); Donald Michie, "Current Developments in Artificial Intelligence and Expert Systems," *Zygon* 20 (1985), pp. 375–90.

6. Ibid., p. 11.

7. Ibid.
8. Ibid., p. xvi. See also Theodore Roszak's recent discussion on the computer and the commodification of knowledge in *The Cult of Information* (New York: Pantheon, 1986).
9. Yoneji Masuda, *The Information Society* (Washington, D.C.: World Future Society, 1980), p. 3.
10. Ibid., p. 148.
11. Ibid.
12. Ibid., p. 74.
13. Ibid., p. 71.
14. Turkle, *The Second Self*, p. 307.
15. Ibid.
16. Bolter, *Turing's Man*, p. 224.
17. Ibid., p. 225. Cf. Herbert Dreyfus, *What Computers Can't Do: A Critique of Artificial Reason* (New York: Harper & Row, 1972).
18. Eric Jantsch, *The Self-Organizing Universe: Scientific and Human Implications of the Emerging Paradigm of Evolution* (New York: Pergamon Press, 1980), p. 263.
19. Ibid., p. 265.
20. Ibid., p. 271.
21. Ibid. Consider in this same context Michel Foucault's critique of the politics of "bio-power" in the power-knowledge dialectic. Foucault, *Power/Knowledge: Selected Interviews and Other Writings 1972–1977* (New York: Pantheon, 1980), pp. 55–62; Hubert L. Dreyfus and Paul Rabinow, *Michel Foucault: Beyond Structuralism and Hermeneutics* (Chicago: The University of Chicago Press, 1982), pp. 126–42.
22. Masuda, *The Information Society*, p. 150.
23. Jeremy Rifkin, *Algeny* (New York: Viking Press, 1983), p. 225.
24. Quoted in Joseph Weizenbaum, *Computer Power and Human Reason* (San Francisco: W. H. Freeman and Co., 1976), p. 247.
25. Edward Feigenbaum, *The Fifth Generation*, p. 251.
26. Masuda, *The Information Society*, p. 3.
27. Weizenbaum, *Computer Power and Human Reason*, p. 115.
28. Author interview, Rockport, Mass., 25 July 1985.
29. Craig Brod, *Technostress* (Reading, Mass.: Addison-Wesley, 1984), pp. 39–40.
30. David Burnham, *The Rise of the Computer State* (New York: Random House, 1982). See also Burnham, "Data Protection," in *The Information Technology Revolution*, ed. Tom Forester (Cambridge: MIT Press, 1985), p. 546.
31. Burnham, "Data Protection," pp. 546–47.
32. Ibid., p. 547.

33. Ibid., p. 548.

34. Ibid.

35. Alvin Toffler, *The Third Wave* (New York: Bantam Books, 1980), p. 320.

36. Ibid.

37. Howard Perlmutter, "Supergiant Firms in the Future," *Wharton Quarterly* (Winter 1968).

38. Statistic from Tudd Polk, senior economist at the U. S. Chamber of Commerce. Quoted in Richard Barriet and Ronald E. Mueller, *Global Reach* (New York: Simon & Schuster, 1975), p. 26.

39. *World Almanac 1975* (New York: Newspaper Enterprise Association, 1974), p. 587. Council on Economic Priorities, *Guide to Corporations: A Social Perspective* (Chicago: Swallow Press, 1974), pp. 2–3.

40. Jerry Mander, "Six Grave Doubts About Computers," *Whole Earth Review* 44 (1985), p. 20.

41. Ibid.

42. Ibid. See also Langdon Winner, "Mythinformation," *Whole Earth Review* 44 (1985), p. 28.

12. TIME PYRAMIDS AND TIME GHETTOS

1. Kurt Lewin, "Time Perspective and Morale," in *Civilian Morale,* ed. Goodwyn B. Watson (New York: Cornwall, 1942), p. 63.

2. Lawrence LeShan, "Time Orientation and Social Class," *Journal of Abnormal Psychology* 47 (1952), p. 589.

3. Ibid.

4. Ibid.

5. Ibid., pp. 590–91.

6. Ibid., p. 591.

7. H. Nowotny, "Time Structuring and Time Measurement: On the Interrelation Between Time-Keepers and Social Time," in *The Study of Time II,* eds. J. T. Fraser and N. Lawrence (New York: Springer-Verlag, 1975), p. 328. See also D. Lewis, *Five Families: Mexican Case Studies in the Culture of Poverty* (New York: Basic Books, 1959); S. M. Miller, F. Riesmann, A. A. Seagull, "Poverty and Self-Indulgence: A Critique of the Non-Deferred Gratification Pattern," in *Poverty in America,* eds. L. A. Forman, J. L. Kornbluth, and A. Haber (Ann Arbor: The University of Michigan Press, 1965).

8. Edward C. Banfield, *The Unheavenly City: The Nature and Future of Our Urban Crisis* (Boston: Little, Brown, 1968), pp. 125–26.

9. Elliot Liebow, *Tally's Corner: A Study of Negro Streetcorner Men*

(Boston: Little, Brown, 1967), p. 65. See also Thomas J. Cottle and Stephen L. Klineberg, *The Present of Things Future* (New York: The Free Press/Macmillan, 1974), pp. 187–88.

10. Lewis Mumford, *The Myth of the Machine,* vol. 1 (New York: Harcourt, Brace and World, 1967), pp. 140–41.

11. See Douglas Noble, "Computer Literacy and Ideology," *Teachers College Record* 85 (1984), pp. 602–14; Joseph Menosky, "Computer Literacy and the Press," *Teachers College Record* 85 (1984), pp. 615–21. For a critique of the modern fetish for information and computer literacy, see Henryk Skolimowski, "Information—Yes, But Where Has All Our Wisdom Gone?" *The Ecologist* 14 (1984), pp. 232–34.

PART IV / COSMIC TIMEPIECES AND POLITICAL LEGITIMACY

13. THE CLOCKWORK UNIVERSE

1. Daniel J. Boorstin, *The Discoverers* (New York: Random House, 1983), p. 71.

2. Samuel L. Macey, *Clocks and the Cosmos: Time in Western Life and Thought* (Hamden, Conn.: Archon Books, 1980), p. 73.

3. Quoted in ibid., p. 107.

4. Ibid.

5. Quoted in ibid., p. 74.

6. Quoted in ibid., p. 76. Descartes would also use the same analogy for his mechanical universe. In his *Principles,* Descartes says that he has "described the earth, and all the world that is visible, as if it were simply a machine. . . ." (4.228). For a stimulating critique of Descartes's "clockwork" view of nature and animals, see Stephen Walker, *Animal Thought* (London: Routledge and Kegan Paul, 1983).

7. Boorstin, *The Discoverers,* pp. 71–72.

8. A. Cornelius Benjamin, "Ideas of Time in the History of Philosophy," in *The Voices of Time,* ed. J. T. Fraser (Amherst: The University of Massachusetts Press, 1981), p. 18.

9. These are the acrid words of Welsh bard Dafydd ap Gwvilyn as quoted in Lawrence Wright, *Clockwork Man* (New York: Horizon Press, 1969), p. 68.

10. Quoted in Macey, *Clocks and the Cosmos,* p. 155.

11. Ibid., p. 162.

12. Quoted in ibid., p. 134.

14. THE INFORMATION UNIVERSE

1. See, for example, Jeremy Campbell, *Grammatical Man: Information, Entropy, Language, and Life* (New York: Simon & Schuster, 1982). One could easily argue that the contemporary "linguistic turn" in philosophy (for example, the work of Derrida, Habermas, Barthes) is but an intellectual by-product of the contemporary interest in information and communications theory. For a powerful critique of semiotics, its relationship to language theory, and the possible role such disciplines play in the social milieu, see J. Ransdell, "Semiotics and Linguistics," in *The Signifying Animal*, eds. I. Rauch and G. F. Carr (Bloomington: The University of Indiana Press, 1980), pp. 135–185. See also William Barrett, *Death of the Soul: From Descartes to the Computer* (New York: Anchor/Doubleday, 1986); Walter J. Ong, *Interfaces of the Word* (Ithaca, N.Y.: Cornell University Press, 1977). Jurgen Habermas's paralinguistic attempt to deconstruct modernity via the "speech act" is a good example of the pragmatism found in present philosophical discourse. This approach, as David Held and others have argued, all but neglects precognitive or preverbal dimensions of consciousness. See Henning Ottman, "Cognitive Interests and Self-Reflection," in *Habermas: Critical Debates,* eds. John B. Thompson and David Held (Cambridge: MIT Press, 1982), pp. 79–97.

2. Campbell, pp. 73–74, 253.

3. Norbert Wiener, *The Human Use of Human Beings* (New York: Avon Books, 1954), p. 278.

4. Ibid.

5. William H. Thorpe and Oliver L. Zangwill, *Current Problems in Animal Behavior* (Cambridge: At the University Press, 1980), p. 303.

6. William H. Thorpe et al., "The Frontiers of Biology," in *Mind in Nature,* eds. John B. Cobb and David Ray Griffin (Washington, D.C.: The University Press of America, 1977), p. 3.

7. Ibid., p. 6.

8. Kenneth Sayre, *Cybernetics and the Philosophy of Mind* (Highlands, N.J.: Humanities Press, 1976), p. xi.

9. Pierre-P. Grassé, *Evolution of Living Organisms: Evidence for a New Theory of Transformation* (New York: Academic Press, 1977), pp. 223–26.

10. Sayre, *Cybernetics and the Philosophy of Mind*, p. 231.

11. George A. Miller, "Language, Learning, and Models of the Mind," unpublished manuscript, June 1972.

12. Geoff Simons, *Silicon Shock: The Menace of the Computer* (Oxford: Basil Blackwell, 1985), p. 9.

13. Sherry Turkle, *The Second Self: Computers and the Human Spirit* (New York: Simon & Schuster, 1984), p. 309.

14. Ibid., p. 17. See also Donald M. MacKay, "Machines, Brains, and Persons," *Zygon* 20 (1985), pp. 401–12.

15. Ibid., p. 289.

16. Ibid., p. 277.

17. Quoted in ibid., p. 288.

18. Ibid.

19. Quoted in Craig Brod, *Technostress* (Reading, Mass.: Addison-Wesley, 1984), p. 10.

20. Quoted in Turkle, *The Second Self*, p. 288.

PART V / THE DEMOCRATIZATION OF TIME

16. BEYOND LEFT AND RIGHT

1. For a general survey of literature on "deep ecology" stewardship, see Bill Devall and George Sessions, *Deep Ecology: Living As If Nature Matters* (Salt Lake City: Peregrine Smith Books, 1985); Michael Tobias, ed., *Deep Ecology* (San Diego: Avant Books, 1985). Cf. Erazim Kohák, *The Embers and the Stars: A Philosophical Inquiry into the Moral Sense of Nature* (Chicago: The University of Chicago Press, 1984); Neil Evernden, *The Natural Alien: Humankind and Environment* (Toronto: The University of Toronto Press, 1985). For the Christian stewardship position, see Wesley Granberg-Michaelson, *A Worldly Spirituality: The Call to Take Care of the Earth* (San Francisco: Harper & Row, 1984); John Carmody, *Ecology and Religion: Toward a New Christian Theology of Nature* (New York: Paulist Press, 1983); Loren Wilkinson, ed., *Earthkeeping* (Grand Rapids, Mich.: William B. Eerdmans, 1980); Peter Reinhart, ed., *To Be Christian Is to Be Ecologist* (San Francisco: Epiphany Press, 1985).

2. Harris Poll, *Washington Post*, 23 May 1977.

SELECTED
BIBLIOGRAPHY

BOOKS

Ayensu, Edward S., and Whitfield, Philip. *The Rhythms of Life*. New York: Crown Publishers, 1982.

Bellah, Robert N. *Tokugawa Religion*. Glencoe, Ill.: The Free Press, 1957.

Bentov, Itzhak. *Stalking the Wild Pendulum*. New York: Bantam Books, 1981.

Berger, Peter L., and Luckmann, Thomas. *The Social Construction of Reality*. New York: Anchor/Doubleday, 1967.

Bergson, Henri Louis. *Time and Free Will*. New York: The Macmillan Company, 1950.

Boden, Margaret A. *Artificial Intelligence and Natural Man*. New York: Basic Books, 1977.

Bolter, J. David. *Turing's Man: Western Civilization in the Computer Age*. Chapel Hill: The University of North Carolina Press, 1984.

Boorstin, Daniel J. *The Discoverers*. New York: Random House, 1983.

Bradley, Michael. *The Chronos Complex I*. Ontario, Canada: Nelson, Foster & Scott, 1973.

Brandon, S. G. F. *History, Time and Deity*. New York: Barnes & Noble, 1965.

Braverman, Harry. *Labor and Monopoly Capital*. New York: Monthly Review Press, 1974.

Brod, Craig. *Technostress: The Human Cost of the Computer Revolution*. Reading, Mass.: Addison-Wesley, 1984.

Brown, N. O. *Life Against Death*. New York: Vintage Books, 1959.

Burnham, David. *The Rise of the Computer State*. New York: Random House, 1980.

Campbell, Jeremy. *Grammatical Man: Information, Entropy, Language and Life*. New York: Simon & Schuster, 1982.

Capra, Fritjof, and Spretnak, Charlene. *Green Politics: The Global Promise*. New York: E. P. Dutton, 1984.

Carlstein, Tommy. *Time Resources, Society, and Ecology*. Boston: Allen & Unwin, 1982.

Carlstein, Tommy, et al. *Making Sense of Time*. London: E. Arnold, 1978.

Claugh, Mary Frances. *Time and Its Importance in Modern Thought*. New York: Russell & Russell, 1970.

Cohen, John. *Psychological Time in Health and Disease*. Springfield, Ill.: Charles C. Thomas, 1967.

Cohn, Norman. *The Pursuit of the Millennium*. London: Temple Smith, 1957.

Cottle, Thomas J., and Klineberg, Stephen. *The Present of Things Future*. New York: The Free Press/Macmillan, 1974.

Darby, Thomas. *The Feast: Meditations on Politics and Time*. Ontario, Canada: The University of Toronto Press, 1982.

De Grazia, Sebastian. *Of Time, Work, and Leisure*. New York: The Twentieth Century Fund, 1962.

Deloria, Vine Jr. *The Metaphysics of Modern Existence*. San Francisco: Harper & Row, 1979.

Devall, Bill, and Sessions, George. *Deep Ecology: Living as if Nature Matters*. Salt Lake City: Peregrine Smith, 1985.

Doob, Leonard W. *The Pattering of Time*. New Haven: Yale University Press, 1971.

Dossey, Larry. *Space, Time and Medicine*. Boulder, Colo.: Shambhala Press, 1982.

Durkheim, Emile. *The Elementary Forms of Religious Life*. New York: The Free Press, 1947.

Edlund, Matthew. *Psychological Time and Mental Illness*. New York: Gardner Press, 1986.

Eliade, Mircea. *The Myth of the Eternal Return*. Princeton, N.J.: Princeton University Press, 1954.

———. *The Sacred and the Profane*. New York: Harcourt, Brace and World, 1959.

Eliot, Jacques. *The Form of Time*. New York: Crane Russak, 1982.

Elkind, David. *The Child and Society: Essays in Applied Child Development*. New York: Oxford University Press, 1979.

Ellul, Jacques. *The Technological Society*. New York: Vintage Books, 1964.

Elvee, Richard Q., ed. *Mind in Nature*. San Francisco: Harper & Row, 1982.

Feigenbaum, Edward A., and McCorduck, Pamela. *The Fifth Generation: Artificial Intelligence and Japan's Challenge to the World*. New York: New American Library, 1984.

Fisher, Helen E. *The Sex Contract: The Evolution of Human Behavior*. New York: William Morrow, 1982.

Forester, Tom, ed. *The Information Technology Revolution*. Cambridge: MIT Press, 1985.

Fortes, Meyer. *Time and Social Structure and Other Essays*. New York: Humanities Press, 1970.

Fraisse, Paul. *The Psychology of Time*. New York: Harper & Row, 1963.

Fraser, J. T. *Of Time, Passion, and Knowledge*. New York: George Braziller, 1975.

———. *The Voices of Time*. Amherst: The University of Massachusetts Press, 1981.

Fraser, J. T., Haber, F. C., and Muller, G. H., eds. *The Study of Time*. New York: Springer-Verlag, 1972.

Fraser, J. T., and Lawrence, N., eds. *The Study of Time II*. New York: Springer-Verlag, 1975.

Fraser, J. T., Lawrence, N., and Park, D., eds. *The Study of Time III*. New York: Springer-Verlag, 1978.

———. *The Study of Time IV*. New York: Springer-Verlag, 1981.

Freud, Sigmund. *Civilization and Its Discontents*. London: Hogarth Press, 1930.

Gardet, L. *Cultures and Time*. Paris: The Unesco Press, 1976.

Gimpel, Jean. *The Medieval Machine*. New York: Holt, Rinehart and Winston, 1976.

Guitton, Jean. *Man in Time*. Notre Dame, Ind.: University of Notre Dame Press, 1966.

Gunnell, Jon G. *Political Philosophy and Time*. Middletown, Conn.: Wesleyan University Press, 1968.

Gurvitch, Georges. *The Spectrum of Social Time*. Dordrecht, Holland: D. Reidel, 1964.

Hall, Edward T. *The Dance of Life: The Other Dimension of Time*. New York: Anchor/Doubleday, 1983.

Hallowell, John. *Main Currents in Modern Political Thought*. New York: Holt, Rinehart and Winston, 1950.

Hawley, Amos. *Human Ecology*. New York: Ronald Press, 1950.

Heidegger, Martin. *Being and Time*. New York: Harper & Row, 1962.

Huxley, Aldous. *The Perennial Philosophy*. San Francisco: Harper & Row, 1970.

Jantsch, Erich. *The Self-Organizing Universe*. New York: Pergamon Press, 1984.

Jonas, Hans. *The Phenomenon of Life*. Chicago: The University of Chicago Press, 1966.

Jung, Carl G. *Psychological Types*. London: Pantheon Books, 1923.

Kantor, D., and Lehr, W. *Inside the Family*. San Francisco: Jossey-Bass Publishers, 1975.

Kern, Stephen. *The Culture of Time and Space, 1880–1918*. Cambridge: Harvard University Press, 1983.

Kidder, Tracy. *The Soul of a New Machine*. Boston: Little, Brown, 1981.

Korzybski, A. *Time Binding: The General Theory*. Lancaster, Pa.: The International Non-Aristotelian Library, 1956.

Landes, David S. *Revolution in Time*. Cambridge: Harvard University Press, 1983.

Le Goff, Jacques. *Time, Work, and Culture in the Middle Ages*. Chicago: The University of Chicago Press, 1980.

Leonard, George B. *The Silent Pulse*. New York: E. P. Dutton, 1978.

Lipset, Seymour M. *Political Man: The Social Bases of Politics*. New York: Anchor/Doubleday, 1963.

Lovejoy, Arthur O. *The Great Chain of Being*. Cambridge, Mass.: Oxford University Press, 1936.

Lovelock, J. E. *Gaia: A New Look at Life on Earth*. Oxford: Oxford University Press, 1979.

Luce, Gay Gaer. *Biological Rhythms in Human and Animal Physiology*. New York: Dover Publications, 1971.

———. *Body Time*. London: Temple Smith, 1972.

Macey, Samuel L. *Clocks and the Cosmos*. Hamden, Conn.: Archon Books, 1980.

Marcuse, Herbert. *Eros and Civilization*. New York: Vintage Books, 1962.

Masuda, Yoneji. *The Information Society*, Washington, D.C.: World Future Society, 1980.

McLuhan, Marshall. *Understanding Media: The Extensions of Man*. New York: McGraw-Hill, 1965.

McNeill, William. *The Rise of the West: A History of the Human Community*. Chicago: The University of Chicago Press, 1963.

———. *The Pursuit of Power*. Chicago: The University of Chicago Press, 1982.

Meerloo, Joost A. M. *Along the Fourth Dimension*. New York: The John Day Co., 1970.

Melges, Frederick Towne. *Time and the Inner Future*. New York: John Wiley & Sons, 1982.

Moore, Wilbert E. *Man, Time, and Society*. New York: John Wiley & Sons, 1963.

Moore-Ede, Martin C., Sulzman, Frank M., and Fuller, Charles A. *The Clocks That Time Us*. Cambridge: Harvard University Press, 1982.

Morris, Richard. *Scientific Attitudes Toward Time*. New York: Simon & Schuster, 1984.

Mumford, Lewis. *Technics and Civilization*. New York: Harcourt Brace & Jovanovich, 1962.

———. *The Myth of the Machine: Technics and Human Development*. New York: Harcourt Brace & Jovanovich, 1967.

———. *The Culture of Cities*. New York: Harcourt Brace & Jovanovich, 1970.

Nora, Simon, and Minc, Alain. *The Computerization of Society*. Cambridge: MIT Press, 1981.

Ong, Walter J. *Orality and Literacy: The Technologizing of the Word*. New York: Methuen & Co., 1982.

Ornstein, Robert E. *On the Experience of Time*. Harmondsworth, England: Penguin Books, 1969.

———. *The Psychology of Consciousness*. New York: W. W. Freeman & Co., 1972.

Papert, Seymour. *Children, Computers, and Powerful Ideas*. New York: Basic Books, 1980.

Park, David. *The Image of Eternity: Roots of Time in the Physical World*. Amherst: The University of Massachusetts Press, 1980.

Piaget, Jean. *The Child's Conception of Time*. New York: Basic Books, 1969.

Polak, Frederick L. *The Image of the Future*. Vols. 1, 2. Leyden, Netherlands: A. W. Sijthoff, 1961.

Pollard, Sidney. *The Genesis of Modern Management*. Cambridge: Harvard University Press, 1965.

Priestley, J. B. *Man and Time*. New York: Dell Publishing Co., 1968.

Prigogine, Ilya, and Stengers, Isabelle. *Order out of Chaos: Man's New Dialogue with Nature*. New York: Bantam Books, 1984.

Quinones, Ricardo J. *The Renaissance Discovery of Time*. Cambridge: Harvard University Press, 1972.

Rawlence, Christopher, ed. *About Time*. London: Jonathan Cape, 1985.

Ricoeur, Paul. *Time and Narrative.* Vol. 1. Chicago: The University of Chicago Press, 1984.

Rifkin, Jeremy. *Declaration of a Heretic.* Boston: Routledge and Kegan Paul, 1985.

Rifkin, Jeremy, with Howard, Ted. *The Emerging Order: God in the Age of Scarcity.* New York: Ballantine Books, 1979.

——. *Entropy: A New World View.* New York: Bantam Books, 1981.

Rifkin, Jeremy, in collaboration with Perlas, Nicanor. *Algeny.* New York: Penguin Books, 1984.

Rokeach, Milton, ed. *The Open and Closed Mind.* New York: Basic Books, 1960.

Sayre, Kenneth. *Cybernetics and the Philosophy of Mind.* Highlands, N.J.: Humanities Press, 1976.

Sherover, Charles M. *The Human Experience of Time.* New York: New York University Press, 1975.

Simon, Herbert A. *The Sciences of the Artificial.* Cambridge: MIT Press, 1969.

Simons, Geoff. *Silicon Shock: The Menace of the Computer Invasion.* New York: Basil Blackwell, 1985.

Smith, Adam. *An Inquiry into the Nature and Cause of the Wealth of Nations.* Edited by Edwin Cannon. London: Methuen & Co., 1961.

Sorokin, Pitirim A. *Sociocultural Causality, Space, Time.* New York: Russell & Russell, 1964.

Stace, W. T. *Time and Eternity.* Princeton, N.J.: Princeton University Press, 1952.

Steiner, Rudolph. *The Riddles of Philosophy.* New York: The Anthroposophic Press, 1973.

Taylor, Frederick. *The Principles of Scientific Management.* New York: W. W. Norton, 1947.

Thompson, William Irving. *The Time Falling Bodies Take to Light.* New York: St. Martin's Press, 1981.

Tobias, Michael, ed. *Deep Ecology.* San Diego: Avant Books, 1985.

Trivers, Howard. *The Rhythm of Being.* New York: Philosophic Library, 1985.

Turkle, Sherry. *The Second Self: Computers and the Human Spirit.* New York: Simon & Schuster, 1984.

Walker, Charles R. *Modern Technology and Civilization.* New York: McGraw-Hill, 1962.

Ward, Ritchie R. *The Living Clocks.* New York: Alfred A. Knopf, 1971.

Weber, Max. *The Protestant Ethic and the Spirit of Capitalism.* New York: Charles Scribner's Sons, 1958.

Weizenbaum, Joseph. *Computer Power and Human Reason.* San Francisco: W. H. Freeman & Co., 1976.

Weizsacker, von, Carl F. *The Unity of Nature.* New York: Farrar, Straus, Giroux, 1980.

Whitrow, G. J. *The Natural Philosophy of Time.* Oxford: Oxford University Press, 1980.

Wiener, Norbert. *The Human Use of Human Beings.* New York: Avon Books, 1967.

Wilson, Peter J. *Man the Promising Primate.* New Haven: Yale University Press, 1980.

Wright, Lawrence. *Clockwork Man.* New York: Horizon Press, 1969.

Yaker, Henri; Osmond, Humphrey; and Cheek, Frances, eds. *The Future of Time: Man's Temporal Environment.* New York: Anchor/Doubleday, 1972.

Zerubavel, Eviatar. *Patterns of Time in Hospital Life: The Temporal Structure of Organization.* Chicago: The University of Chicago Press, 1979.

———. *Hidden Rhythms: Schedules and Calendars in Social Life.* Chicago: The University of Chicago Press, 1981.

———. *The Seven Day Circle.* New York: The Free Press/Macmillan, 1985.

Zubin, J., and Hunt, H. F., eds. *Comparative Psychopathology.* New York: Grune & Stratton, 1967.

ARTICLES

Ames, Louise Bates. "The Development of the Sense of Time in the Young Child." *Journal of Genetic Psychology* 68 (1946).

Ariotti, Piero E. "The Conception of Time in Late Antiquity." *International Philosophical Quarterly* 12 (1972).

Aschoff, Jurgen. "Circadian Rhythms in Man." *Science* 148 (1965).

Baldwin, Robert O. "Sex Differences in the Sense of Time: Failure to Replicate a 1904 Study." *Perceptual and Motor Skills* 22 (1966).

Banks, Robin, and Cappon, Daniel. "Effect of Reduced Sensory Input on Time Perception." *Perceptual and Motor Skills* 14 (1962).

Barabasz, Arreed. "Time Estimation and Temporal Orientation in Delinquents and Nondelinquents: A Re-examination." *Journal of General Psychology* 82 (1970).

———. "Temporal Orientation and Academic Achievement in College." *Journal of Social Psychology* 80 (1970).

Barndt, Robert J., and Johnson, Donald M. "Time Orientation in Delinquents." *Journal of Abnormal and Social Psychology* 51 (1955).

Barton, K., and Cattel, R. B. "Changes in Psychological State Measures and Time of Day." *Psychological Reports* 35 (1974).

Beilin, Harry. "The Pattern of Postponability and Its Relation to Social Class Mobility." *The Journal of Social Psychology* 44 (1956).

Bell, Daniel. "The Clock Watchers: Americans at Work." *Time,* 8 September 1975.

Bell, C. R., and Watts, Anne N. "Personality and Judgements of Temporal Intervals." *British Journal of Psychology* 57 (1966).

Bell, C. R., and Provins, K. A. "Relation between Physiological Responses to Environmental Heat and Time Judgments." *Journal of Experimental Psychology* 6 (1963).

Bell, P. M. "Sense of Time." *New Science* 15 (1975).

Bell, Wendell, and Mau, James A. "Images of the Future: Theory and Research Strategies." *Theoretical Sociology: Perspectives and Development.* Edited by John C. McKinney and Edward A. Tiryakian. New York: Appleton-Century-Crofts-Meredith, 1970.

Bergler, Edmund, and Roheim, Geza. "Psychology of Time Perception." *Psychoanalytic Quarterly* 15 (1946).

Berman, Morris. "The Cybernetic Dream of the Twenty-first Century." *Journal of Humanistic Psychology* 26 (1986).

Berndt, Thomas J., and Wood, David J. "The Development of Time Concepts through Conflict Based on a Primitive Duration Capacity." *Child Development* 45 (1974).

Berstein, Basil. "Language and Social Class." *British Journal of Sociology* 2 (1960).

Bettelheim, Bruno. "Individual and Mass Behavior in Extreme Situations." *Journal of Abnormal and Social Psychology* 38 (1943).

Blake, M. J. F. "Relationship Between Circadian Rhythm of Body Temperature and Introversion-Extraversion." *Nature,* 19 August 1967.

Bonaparte, Marie. "Time and the Unconscious." *International Journal of Psycho-Analysis* 21 (1940).

Bouwsma, William J. "Anxiety and the Formation of Early Modern Culture." *After the Reformation.* Edited by Barbara Coralament. Philadelphia: The University of Pennsylvania Press, 1980.

Bradley, N. C. "The Growth of the Knowledge of Time in Children of School-Age." *British Journal of Psychology* 38 (1947).

Bradley, Warren M. "The Clock Manifesto." *New York Academy of Sciences Perspectives of Time* (1966).

Bregman, Lucy. "Growing Older Together: Temporality, Mutuality,

and Performance in the Thought of Alfred Schutz and Erik Erikson."
The Journal of Religion 53 (April 1973).

Brock, Timothy C., and Guidice, Carolyn D. "Stealing and Temporal
Orientation." *Journal of Abnormal and Social Psychology* 66 (1963).

Brody, Jane E. "The Varied Times of Our Lives: How Chronobiology
Affects Our Bodies." *New York Times,* 15 August 1984.

Brown, D. G. "The Value of Time." *Ethics: An International Journal
of Social, Political, and Legal Philosophy* 80 (April 1970).

Brown, Frank A. "The Rhythmic Nature of Animals and Plants."
American Scientists 47 (June 1959).

Bull, Neil C. "Chronology: The Field of Social Time." *Journal of Leisure Research* 10 (1978).

Burman, Rickie. "Time and Socioeconomic Change on Simbo Solomon Islands." *Man: The Journal of the Royal Anthropological Institute*
16 (1981).

Calkins, Kathy. "Time: Perspectives, Marking and Styles of Usage."
Social Problems 17 (1970).

Campos, Leonard P. "Relationship Between Time Estimation and Retentive Personality Traits." *Perceptual and Motor Skills* 23 (1966).

Chessick, Richard D. "The Sense of Reality, Time, and Creative Inspiration." *American Imago* 14 (1957).

Condon, Willam S. "A Primary Phase in the Organization of Infant
Responding Behavior." *Studies in Mother-Infant Interaction.* Edited by
H. R. Schaffer. New York: Academic Press, 1977.

———. "Neonate Movement Is Synchronized with Adult Speech: Interactional Participation and Language Acquisition." *Science* 103 (1974).

———. "Multiple Response to Sound in Dysfunctional Children."
Journal of Autism and Childhood Schizophrenia 5 (1975).

Condon, W. S., and Ogston, W. D. "Sound Film Analysis of Normal
and Pathological Behavior Patterns." *The Journal of Nervous and Mental
Disease* 143 (1966).

———. "A Method of Studying Animal Behavior." *The Journal of
Auditory Research* 7 (1967).

Coser, L. A., and Coser, R. L. "Time and Social Structure." *Modern
Sociology.* Edited by A. W. Gouldner and H. P. Gouldner. New York:
Harcourt, Brace, and World, 1963.

Cottle, Thomas J. "Future Orientations and Avoidance: Speculations on
the Time of Achievement." *The Journal of Social Psychology* 67 (1965).

———. "The Circles Test: An Investigation of Perceptions of Temporal
Relatedness and Dominance." *Journal of Projective Techniques and
Personality Assessment* 31 (1967).

Cromwell, Ronald E.; Keeney, Bradford P.; and Adams, Bert W. "Temporal Patterning in the Family." *Family Process* 15 (1976).

Cuffaro, Harriet K. "Microcomputer in Education: Why Is Earlier Better?" *Teachers College Record* 85 (Summer 1984).

Danzger, Herbert M. "Community Power Structure: Problems and Continuities." *American Sociological Review* 29 (1964).

Darnley, Fred. "Periodicity in the Family: What Is It and How Does It Work?" *Family Relations* (January 1981).

Davids, A., Kidder, C., and Reich, M. "Time Orientation in Male and Female Juvenile Delinquents." *Journal of Abnormal and Social Psychology* 64 (1962).

Davids, A., and Parenti, A. W. "Time Orientation and Interpersonal Relations of Emotionally Disturbed and Normal Children." *Journal of Abnormal and Social Psychology* 57 (1958).

Davidson, Graham R., and Klich, Leon Z. "Cultural Factors in the Development of Temporal and Spatial Ordering." *Child Development* 51 (1980).

Davy, John. "Mindstorms in the Lamplight." *Teachers College Record* 85 (Summer 1984).

Dossey, Larry. "Medicine Enters the Fourth Dimension." *Science Digest* 90 (October 1982).

Douvan, Elizabeth, and Adelson, Joseph. "The Psychodynamics of Social Mobility in Adolescent Boys." *Journal of Abnormal and Social Psychology* 56 (1958).

Dubos, René; Goodman, Paul, et al. "The Fitness of Man's Environment." Papers delivered at the Smithsonian Institution Annual Symposium, Washington, D.C., 16–18 February 1967.

Du Preez, P. D. "Judgment of Time and Aspects of Personality." *Journal of Abnormal and Social Psychology* 69 (1964).

Earman, John. "An Attempt to Add a Little Direction to 'The Problem of the Direction of Time.' " *Philosophy of Science* 41 (1974).

Ellis, Laura M., et al. "Time Orientation and Social Class: An Experimental Supplement." *Journal of Abnormal and Social Psychology* 51 (1955).

Ellis, Robert A.; Lane, Clayton W., et al. "The Index of Class Position: An Improved Intercommunity Measure of Stratification." *American Sociological Review* 28 (1963).

Engel-Frisch, Gladys. "Some Neglected Temporal Aspects of Human Ecology." *Social Forces* 22 (1943).

Essman, Walter B. "Temporal Problem Solving." *Perceptual and Motor Skills* 8 (1958).

Falk, John L., and Bindra, Dalbir. "Judgment of Time As a Function of Serial Position and Stress." *Journal of Experimental Psychology* 47 (1954).

Farber, Maurice L. "The Armageddon Complex: Dynamics of Opinion." *Public Opinion Quarterly* 15 (1951).

———. "Time-Perspective and Feeling-Tone: A Study in the Perception of Days." *Journal of Psychology* 35 (1953).

Floyd, Keith. "Of Time and Mind: From Paradox to Paradigm." *Frontiers of Consciousness.* Edited by John White. New York: Julian Press, 1974.

Germain, Carel B. "Time: An Ecological Variable in Social Work Practice." *Social Casework* 57 (1976).

Goldfarb, William. "Psychological Privation in Infancy and Subsequent Adjustment." *American Journal of Orthopsychiatry* (1945).

Goode, William J. "A Theory of Role Strain." *American Sociological Review* 25 (1960).

Goodin, Robert E. "Banana Time in British Politics." *Political Studies* 30 (1982).

Goodman, Richard A. "Environmental Knowledge and Organizational Time Horizon: Some Functions and Dysfunctions." *Human Relations* 26 (1973).

Goody, Jack. "Time: Social Organization." *The International Encyclopedia of the Social Sciences.* Vol. 16. Edited by David L. Sills. New York: Macmillan, 1968.

Gurvitch, George. "Social Structure and the Multiplicity of Time." *Sociological Theory, Values, and Sociocultural Change.* Edited by Edward A. Tiryakian. New York: The Free Press of Glencoe, 1963.

Gray, Charles E. "Paradoxes in Western Creativity." *American Anthropologist* (1974).

Halberg, Franz, et al. "Reading, 'Riting, 'Rithmetic—and Rhythms: A New 'Relevant' 'R' in the Educative Process." *Perspectives in Biology and Medicine* 17 (1973).

Hallowell, Irving A. "Temporal Orientation in Western Civilization and in a Pre-literate Society." *American Anthropologist* 39 (1937).

Hareven, Tamara K. "Family Time and Historical Time." *Daedalus* 106 (Spring, 1977).

Henry, Jules. "White People's Time, Colored People's Time." *Transaction* 2 (1965).

Henry, William E. "The Business Executive: The Psychodynamics of a Social Role." *American Journal of Sociology* 54 (1949).

Hilts, Philip. "The Clock Within." *Science,* December 1980.

Horton, John. "Time and Cool People." *Transaction* 4 (1967).

Hough, Walter. "Time-Keeping by Light and Fire." *American Anthropologist* 6 (1893).

Howe, Leopold. "The Social Determination of Knowledge: Maurice Bloch and Balinese Time." *Man: The Journal of the Royal Anthropological Institute* 16 (1981).

Iutcovich, Mark; Babbitt, Charles E.; and Iutcovich, Joyce. "Time Perception: A Case Study of a Developing Nation." *Sociological Focus* 12 (1979).

Jenson, Arthur R. "Political Ideologies and Educational Research." *Phi Delta Kappan*, March 1984.

Kastenbaum, Robert. "The Dimensions of Future Time Perspective, An Experimental Analysis." *Journal of General Psychology* 65 (1961).

———. "The Direction of Time Perception: The Influence of Affective Set." *Journal of General Psychology* 73 (1965).

Keller, Suzanne. "Ambition and Social Class: A Respecification." *Social Forces* 43 (1964).

Klevana, Wanda M. "Does Labor Time Decrease with Industrialization?: A Survey of Time-Allocation Studies." *Current Anthropology* 21 (1980).

Klineberg, S. L. "Future Time Perspective and the Preference for Delayed Reward." *Journal of Personality and Social Psychology* 8 (1968).

Knapp, Robert H., and Garbutt, John T. "Time Imagery and the Achievement Motive." *Journal of Personality* (1958).

———. "Variation in Time Descriptions and Need Achievement." *The Journal of Social Psychology* 67 (1965).

Laver, Robert H. "Temporality and Social Change: The Case Study of 19th Century China and Japan." *Sociological Quarterly* 14 (Autumn 1973).

Leach, E. R. "Primitive Time-Reckoning." *A History of Technology*. Edited by Charles Singer, E. J. Holmyard, and A. R. Hall. Oxford: Clarendon Press, 1954.

———. "Two Essays Concerning the Representation of Time." *Rethinking Anthropology*. New York: Humanities Press, 1966.

Leichter, Hope Jenson. "A Note on Time and Education." *Teachers College Record* 81 (Spring 1980).

LeShan, Lawrence L. "Time Orientation and Social Class." *Journal of Abnormal and Social Psychology* 47 (1957).

Leslie, John. "II: The Value of Time." *American Philosophical Quarterly* 13 (1976).

Lessing, Elise E. "Demographic, Developmental, and Personality Correlates of Length and Future Time Perspective." *Journal of Personality* 36 (1968).

Lewin, Kurt. "Time Perspective and Morale." *Civilian Morale.* Edited by Goodwyn B. Watson. Cornwall, N.Y.: The Cornwall Press, 1942.

Lewis, David J., and Weigert, Andrew J. "The Structures and Meanings of Social Time." *Social Forces* 60 (1981).

Luhmann, Nicklas. "The Future Cannot Begin: Temporal Structure in Modern Society." *Social Research* 43 (Spring 1976).

Lyons, Oren. "An Iroquois Perspective." *American Indian Environments: Ecological Issues in Native American History.* Edited by Christopher Vecsey and Robert W. Venables. New York: Syracuse University Press, 1980.

Mander, Jerry. "Six Grave Doubts About Computers." *Whole Earth Review* 44 (1985).

Marks, Stephen R. "Multiple Roles and Role Strain: Some Notes on Human Energy, Time and Commitment." *American Sociological Review* 42 (1977).

Melbin, Murray. "Behavior Rhythms in Mental Hospitals." *American Journal of Sociology* 74 (1969).

———. "Night as Frontier." *American Sociological Review* 43 (1978).

———. "Settling the Frontier of Night." *Psychology Today,* June 1979.

Michelson, William. "Some Like It Hot: Social Participation and Environmental Use of Functions of Season." *American Journal of Sociology* 76 (1971).

Mills, J. N. "Human Circadian Rhythms." *Physiological Review* 46 (1966).

Mishel, Walter. "Father Absence and Delay of Gratification: Cross Cultural Comparisons." *Journal of Abnormal and Social Psychology* 63 (1961).

Mishel, Walter, and Metzer, Ralph. "Preference for Delayed Reward as a Function of Age, Intelligence, and Length of Delay Interval." *Journal of Abnormal and Social Psychology* 64 (1962).

Montagu, Ashley. "Time-Binding and the Concept of Culture." *Scientific Monthly* 77 (1953).

Moore, Wilbert E. "A Reconsideration of Theories of Social Change." *American Sociological Review* 25 (1960).

———. "Time—Theory, Ultimate Scarcity." *American Behavioral Scientist* 6 (1963).

Noble, Douglas. "Computer Literacy and Ideology." *Teachers College Record* 85 (Summer 1984).

Noyes, Richard. "The Time Horizon of Planned Social Change." Part I. *The American Journal of Economics and Sociology* 39 (1980).

Oberndorf, C. P. "Time—Its Relation to Reality and Purpose." *Psychoanalytic Review* (1941).

O'Rand, Angela, and Ellis, Robert A. "Social Class and Social Time Perspective." *Social Forces* 53 (1974).

Orme, John E. "Time: Psychological Aspects." *Making Sense of Time*. Edited by Tommy Carlstein, Don Parks, and Nigel Thrift. New York: John Wiley & Sons, 1978.

Palmer, John D. "Biorhythm Bunkum." *Natural History* 91 (1982).

Perry, J. A. "Land, Power, and the Lie." *Man: The Journal of the Royal Anthropological Institute* 16 (1981).

Persson, Malin. "Time Perspectives Amongst Criminals." *Acta Sociologica* 24 (1981).

Peterson, Richard A. "Measuring Culture, Leisure, and Time Use." *American Academy of Political and Social Science Annuals* (January 1981).

Platt, J. J., and Taylor, R. E. "Homesickness, Future Time Perspective, and the Self Concept." *Proceeding of the 74th Meeting of the American Psychological Association* (1966).

Pollak, Oliver B. "Efficiency Preparedness and Conservation: The Daylight Savings Time Movement." *History Today* 31 (1981).

Pratt, Lois. "Business Temporal Norms and Bereavement Behavior." *American Sociological Review* 46 (1981).

Price, Marjorie. "Identity through Time." *Journal of Philosophy* 74 (1977).

Putrill, Richard L. "Foreknowledge and Fatalism." *Religious Studies* 10 (1974).

Rabin, Albert I., and Wallace, Melvin. "Temporal Experience." *Psychological Bulletin* 57 (1960).

Rakover, Sam S. "Effect of Discriminative Cues and Time Interval on Tolerance of Pain as a Measure of Fear." *Psychological Reports* 38 (1976).

Rezsöházy, Rudolf. "The Concept of Social Time: Its Role in Development." *International Social Science Journal* (1972).

Roberts, Alan H., and Herrmann, Robert S. "Dogmatism, Time Perspective, and Anomie." *Journal of Individual Psychology* 16 (1959).

Rosen, Bernard C. "The Achievement Syndrome: A Psychocultural Dimension of Social Stratification." *American Sociological Review* 21 (1956).

Ruiz, Rene A. "Tests of Temporal Perspective: Do They Measure the Same Construct?" *Psychological Reports* 21 (1967).

Sardello, Robert J. "The Technological Threat to Education." *Teachers College Record* 85 (Summer 1984).

Schatzman, Leonard, and Strauss, Anselm. "Social Class and Modes of Communication." *American Journal of Sociology* 55 (1955).

Schegloff, Emanuel. "Sequencing in Controversial Openings." *American Anthropologist* 70 (1968).

Schneider, Louis, and Svare, Lysgaard. "The Deferred Gratification Pattern." *American Sociological Review* 18 (1953).

Schwartz, Barry. "Waiting, Exchange, and Power: The Distribution of Time in Social Systems." *American Journal of Sociology* 79 (1974).

———. "The Social Ecology of Time Barriers." *Social Forces* 56 (1978).

———. "Queues, Priorities, and Social Process." *Social Psychology* 41 (1978).

Shaffer, L. H. "Rhythm and Timing in Skill." *Psychological Review* 89 (1982).

Sharp, Sharon. "Biological Rhythms and the Timing of Death." *Omega* 12 (1981–82).

Sivin, N. "Chinese Alchemy and the Manipulation of Time." *Isis* 67 (1976).

Slater, Philip E. "On Social Regression." *American Sociological Review* 28 (1963).

Smith, Marian W. "Different Cultural Concepts of Past, Present, and Future: A Study of Ego Extension." *Psychiatry* 15 (1952).

Smith, Michael French. "Bloody Time and Bloody Scarcity: Capitalism, Authority, and the Transformation of Temporal Experience in a Papua New Guinea Village." *American Ethnologists* (1982).

Sorokin, Pitirim A., and Merton, Robert K. "Social Time: A Methodological and Functional Analysis." *American Journal of Sociology* 42 (1937).

Srole, Leo. "Social Integration and Certain Corollaries: An Exploratory Study." *American Sociological Review* 21 (1956).

Stafford, Frank P. "Women's Use of Time Converging with Men's." *Monthly Labor Review* 103 (1980).

Szalai, Alexander. "Women's Time: Women in the Light of Contemporary Time-Budget Research." *Futures* 7 (1975).

Teahan, John. "Future Time Perspective, Optimism, and Academic Achievement." *Journal of Abnormal and Social Psychology* 57 (1958).

Thomas, Alexander; Chess, Stella; and Birch, Herbert. "The Origin of Personality." *Scientific American* 223 (1970).

Thompson, E. P. "Time, Work-Discipline and Industrial Capitalism." *Past and Present* 38 (1967).

Tickamyer, Ann R. "Wealth and Power: A Comparison of Men and Women in the Property Elite." *The Structure and Meanings of Social Time*. Edited by J. David Lewis and Andrew J. Weigert. Chapel Hill: The University of North Carolina Press, 1981.

Tillman, Mary K. "Temporality and Role-Taking in G. H. Mead." *Social Research* 37 (1970).

Wade, M. G.; Ellis, M. J.; and Bohrer, R. E. "Biorhythms in the Activity of Children During Free Play." *Journal of Experimental Analysis of Behavior* 20 (1973).

Wallace, M., and Rabin, A. I. "Temporal Experience." *Psychological Bulletin* 57 (1960).

Wax, Murray. "The Notions of Nature, Man, and Time of a Hunting People." *Southern Folklore Quarterly* 26 (1962).

Winner, Langdon. "Mythinformation." *Whole Earth Review* 44 (1985).

Woodcock, George. "The Tyranny of the Clock." *Politics* 1 (1944).

Young, Michael, and Ziman, John. "Cycles in Social Behavior." *Nature* 229 (8 January 1971).

Zajonc, Arthur G. "Computer Pedagogy?: Questions Concerning the New Educational Technology." *Teachers College Record* 85 (Summer 1984).

Zerubavel, Eviatar. "The French Republican Calendar: A Case Study in the Sociology of Time." *American Sociological Review* 42 (1977).

———. "Private Time and Public Time: The Temporal Structure of Social Accessibility and Professional Commitments." *Social Forces* 58 (1979).

———. "Easter and Passover: On Calendars and Group Identity." *American Sociological Review* 47 (1982).

Acknowledgments

I would like to thank Donald E. Davis for collaborating on the research that broadened the scope of this book. Andrew Kimbrell provided a useful critique and, together with Nicanor Perlas, helped to shape the final manuscript. Many thoughtful comments were offered by Morris Berman, Laine Shakerdge, Skip Stein, and Donna Wulkan.

My agent, Michael Carlisle, who has believed in this book from the beginning, provided constant encouragement and support. And Jim helped to secure foreign translations of *Time Wars*. Finally, I would like to thank my editor, Jack Macrae III; the structure and presentation of *Time Wars* has truly been a joint effort. I have enjoyed learning from him.

INDEX

251

Horology, 221
Houses, empathetic *vs.* orthodox design of, 202–3
Hruchesky, William, 41
Hunter-gatherers, temporal bond with nature, 70

Images of future, 123–33
Christian, 128–31; *vs.* bourgeois, 134, 137–38, 139
Enlightenment's, 139–42
Jewish, 126–28
materialist, 142–47
simulated, 148–63
Immortality
in computopia, 158
desire for, 124, 125
materialism and, 142
science and technology and, 143
see also Salvation
Imperialism, temporal, 113–17, 144
Indians, American
Iroquois, 65
Pueblo, 49–50, 57
Individualism, *vs.* organic approaches, 63–64, 97
Indonesia, temporal values of, 62
Industrial production, *see* Factories
Information
access to, 161, 170
centralization of, 162
computer transforms time into, 102, 118, 158, 182
energy reduced to, 149
evolution of, *vs.* evolution of consciousness, 153
manipulation of, 149
as organizing principle of new age, 184–88
as power, 162, 170
theft of, 170
Information theory, 181, 183–84, 188

and evolution, 186
and psychology, 186–87
Institute of Chronobiology, New York Hospital, 40
Intelligence
computer as analogy for, 187
linked with rapid learning, 59
simulation of, 149
Intelligence, artificial (AI), 14, 149, 151, 153–54, 222
International Conference on Time (Paris, 1912), 113
International Meridian Conference (1884), 113
Iroquois Nation, time perspective of, 65
Islamic planning devices, 56
Isolation
vs. communion (as modern choice), 193–94
computers and, 19–20
vs. empathy, 195
in workplace, 204
Israel, Jewish calendar in, 71

Jantsch, Eric, 156–57
Japan
computerized economy of, 150, 152–53
computer-run factory in, 99, 101, 222
temporal values of, 61–62, 64; and Western influence, 114, 116
Japan Computer Usage Development Institute, 152
Jesus of Nazareth, 128, 208
emulation of, 130
see also Second Coming of Christ
Jet lag, 37
Jews
calendar of, 71–72

Time perspective
  bourgeois *vs.* workers', 89–97
  computer and, 98–102
  cultural differences in, 64–65
  empathetic *vs.* power dynamic,
    199–200
  and social class, 166–68
Time-sharing, 21
Time systems
  local, 112, 113
  world, standardized, 112–13
Toda, Masanoa, 188
Toffler, Alvin, 149
Tompion, Thomas, 105
Trains, *see* Railroads
Turkle, Sherry, 16, 153–54, 187

Uncertainty
  computer programming "elimi-
    nates," 160
  planning as strategy against, 55,
    138–39
Unconscious, 154
Urban life
  growth of (19th century), 144
  modern temporal values in, 97
Utility of time, 102, 113, 118, 156

Vanderbilt, Amy, 51
Variable speech control, 119
Video games and players, 16

Wagner, Daniel, 40
Wahl, O., 30–31
"Wait state," 21
Waltham Watch Company, 106
Watches
  manufacture of, 105, 106
  stopwatch, for motion-time stud-
    ies, 107
  for trains, 145

Watson, James, 185
Weather forecasting, computers in,
  161
Webster, Hutton: *Rest Days,* 53–
  54
Wedgwood, Josiah, 93
Week, length of, 53
Wizenbaum, Joseph, 161
Whitehead, Alfred North, 2
Whitney, Eli, 105–6
Whitrow, C. J.: *Natural Philosophy
  of Time, The,* 33
Wiener, Norbert, 182, 183
Wilkinson, Robert, 39
Witts and Rodick silk mill, Essex,
  England, 95
Wood, John, 95
Woodcock, George, 80
Workers
  children as, 94–95
  clock regulates, 86
  and computers, 14; psychological
    and behavioral problems, 16–23;
    resistance to, 22–23
  efficiency of, 104–11
  empathetic *vs.* power approach to
    management of, 204–5
  personal satisfaction from work
    desired by, 207
  poor, time skills of, 167
  schedule and, 89–97
  and shift work, 37, 38
  synchronization of, 64
  in third-world countries, time val-
    ues of, 113–14
World time, standardized, 112–13
Wright, Lawrence, 85

Zangwill, Oliver: *Current Problems
  in Animal Behavior,* 184
Zerubavel, Eviatar, 55, 75, 76, 82